Reinventing Modern Dublin

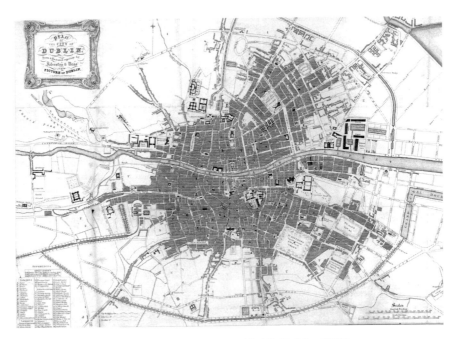

Dublin in 1821, from J. J. McGregor *New picture of Dublin* (Dublin, 1821).

Reinventing Modern Dublin

Streetscape, iconography and the politics of identity

YVONNE WHELAN

UNIVERSITY COLLEGE DUBLIN PRESS
Preas Choláiste Ollscoile Bhaile Átha Cliath

First published 2003
by University College Dublin Press
Newman House
86 St Stephen's Green
Dublin 2
Ireland

www.ucdpress.ie

ISBN 1-900621-85-1 hb
ISBN 1-900621-86-X pb

Cataloguing in Publication data available
from the British Library

Typeset in Ireland in Adobe Garamond, Janson and Trade Gothic
by Elaine Shiels, Bantry, Co. Cork
Text design by Lyn Davies
Printed in England by MPG Books Ltd, Bodmin, Cornwall
This book is printed on acid-free paper

Dedicated to the memory of my parents
Fran and Renee Whelan

Contents

Contents

Plates

Except where otherwise indicated in the text, all illustrations are the author's own

Figures

Tables

Acknowledgements

Books, just like the cultural landscapes in which we live, are products of many stages of development, marked by times of fervent productivity that are all too often tempered by phases of seeming regression. Writing *Reinventing Modern Dublin* has been no different. Its genesis can be traced back to the Geography Department in University College Dublin where my friend and former colleague Professor Anngret Simms supervised the PhD research upon which it is based. I am grateful to Anngret for initially inspiring this work and offering consistent support and wise guidance along the road to publication. In carrying out the research it was my pleasure to encounter a host of people whose assistance I would like to acknowledge. I am grateful to Mary Clark (Dublin City Archives), Valerie Ingram (OPW Library), Máire Kennedy (Gilbert Library) and Gráinne MacLochlainn (National Photographic Archive) for helping me to locate and access material on Dublin, as well as to the staff of the National Library, National Archives and the Irish Architectural Archive. Colleagues at the Department of Geography (UCD), the School of Biological and Environmental Sciences (University of Ulster, Coleraine), and the Academy for Irish Cultural Heritages (University of Ulster), where I am presently based, offered their support and encouragement in many different ways. I am especially grateful to Stephen Hannon (UCD) who gave freely of his time and cartographic expertise in drawing the maps. I would also like to acknowledge the editorial assistance of Barbara Mennell (UCD Press) and Julitta Clancy for compiling the index. The University of Ulster Research Policy and Practice Committee, together with the Academy for Irish Cultural Heritages provided a generous subvention towards the costs of publication and this is gratefully acknowledged. While writing *Reinventing Modern Dublin* many friends and colleagues have offered me their support, advice and encouragement. In particular I would like to thank Stine Berner, Joe Brady, Anne Buttimer, Jonathan Cherry, Michael Dempsey, Brian Graham, Liam Harte, Mike Heffernan, Marjatta Hietala, David Jones, Tim Mooney, Niamh Moore, Máiréad nic Craith, Willie Nolan, Martina O'Donnell, Sinéad Power, Jacinta Prunty and

Acknowledgements

Ann Marie Smith. My deepest debt of gratitude goes to my family in Dublin and especially my parents, Fran and Renee Whelan, who gently fostered a love of learning and provided constant encouragement in my academic endeavours. I dedicate this book to their memory.

Yvonne Whelan
Derry
December 2002

Abbreviations

IAA	Irish Architectural Archive
IRA	Irish Republican Army
IRB	Irish Republican Brotherhood
NLI	National Library of Ireland
OPW	Office of Public Works
RDS	Royal Dublin Society
RPDCD	Reports and Printed Documents of the Corporation of Dublin
TCD	Trinity College, Dublin
UCD	University College Dublin

Setting the context

Towards a reading of Dublin's iconography

On 15 February 1808, crowds gathered in Dublin's city centre to watch as the foundation stone for a monument dedicated to Lord Nelson was laid in a display of pomp and ceremony. While the streets of Dublin were bedecked with bunting and lined with troops from the city garrison, crowds watched as the stone was put in place and gave three cheers to the volleys fired by the Dublin yeomanry. A year later the monument was finished and opened to the public on 21 October 1809, the anniversary of the Battle of Trafalgar. This monument and countless others erected before and after it stood as a symbols of Ireland's status as a city of the British Empire. The scenes that prevailed in 1808, however, stand in marked contrast to events that took place some years later in 1966. In the early morning of 8 March a bomb rocked Dublin's most central thoroughfare. Explosives had been planted by a Republican splinter group at the base of Nelson's Pillar and when the bomb went off the figure of Nelson came toppling to the ground, while the Corinthian column upon which it had stood was virtually destroyed. Engineers from the Irish Army demolished what remained of the pillar in the days following the explosion. Since then the site remained vacant until December 2000 when the Irish Government announced that a new monument known as the 'Dublin Spire' was to be erected on the site of the old pillar. This steel structure, striking for its overtly apolitical nature, is now set to change the iconography of O'Connell Street as well as the entire city.

In many ways the story of Nelson's Pillar captures in microcosm the narrative of Dublin's iconography as it evolved before and after the creation of the Irish Free State in 1922. Before independence it was just one of many monuments, street names and public buildings that represented in concrete form Ireland's place within the United Kingdom of Great Britain and Ireland. They stood as tangible signs of Dublin's position as 'Second City', a phenomenon that was further reinforced by the visits of various members of the British royal family. On these occasions cities and towns throughout the country were lavishly decorated, royal processions through urban spaces were symbolically planned, gigantic temporary structures erected and addresses of welcome carefully crafted. This, however, is only one reading of what had become a much more complex iconography by the late nineteenth century and one that embodied the fraught political context which then prevailed. In fact, various strands of Irish nationalist, republican and socialist opinion used the cultural landscape to express resistance and opposition to empire and to assert that Ireland laboured under a malign form of colonial rule from which every effort should be made to break free.

Reinventing Modern Dublin: streetscape, iconography and the politics of identity takes the reader back to that contested iconography. As a study in historical and cultural geography this work can be situated within a broader range of literature in geography that has highlighted the powerful role of landscape as a site of symbolic representation integral to the imaginative construction of national identity. The book addresses the symbolic geography of Dublin as it evolved in the decades before and after Independence. Attention is focused upon the construction of landscape as a site of signification where issues of naming, building, designing and memorialising became firmly grounded in space and bound up with the representational politics of the day. The capacity of the cultural landscape to underpin, reinforce and legitimate particular narratives of identity is explored through close examination of key landscape elements.

Part I, 'Unravelling the cultural geographies of landscape' establishes the conceptual framework and highlights the key issues that form the focus of subsequent chapters. It begins by addressing the power of landscape as a site of emblematic representation and outlines the developments that have taken place in the field of cultural geography paving the way for such an approach. It is argued that the cultural landscape is integral to processes of nation building and empire formation and that the interplay between landscape and memory is crucial in constructing narratives of identity. The emphasis then shifts towards an exploration of the four specific landscape elements around which the remaining chapters are anchored, namely public statuary, street names, architecture and urban design initiatives. With reference to a range of case studies it is argued that each of these 'icons of identity' contributes a vital layer to the symbolic geography of the city, weaving maps of meaning and representing often conflicting discourses.

In Part II the evolution of Dublin's iconography before the achievement of political independence is examined. It details the processes whereby the urban landscape became highly politicised and contested in a manner that reflected the prevailing political context. While in some respects landscape became a means of cultivating a sense of imperial identity and fostering a feeling of belonging to empire among citizens, it also served as a site upon which to represent resistance. The chapters in this section return to the period before Independence when Dublin, then a deposed capital, stood at an important geographical interface or contact zone between the imperial metropole and the colonised territory. Chapter 2 charts retrospectively the ways in which the 'text' of the city was scripted throughout the eighteenth and nineteenth centuries with reference to public monuments. Particular attention is paid to the role of the symbolic landscape in cultivating a sense of belonging to empire among Dublin's inhabitants through the erection of monuments dedicated to figures of the British monarchy and military and in the pomp and ceremony that went with their unveiling in the city.

In chapter 3 the emergence of contested identities in the monumental landscape becomes the chief focus of attention. This chapter explores the ways in which the city served as a site of resistance and opposition to the colonial 'Other', especially towards the end of the nineteenth century when concerted attempts were made to subvert the inscription of imperial power and to challenge the legitimacy of such a ruling authority. The monuments erected ensured that as Ireland stood on the cusp of independence, its largest city captured in symbolic form the contested nature of the political context. The power of street naming and renaming in contributing to the iconography of Dublin's cultural landscape before 1922 is the chief focus of the fourth chapter. It considers the legacy of street names in the city by focusing especially on the names assigned to the River Liffey's bridges and quays which were broadly representative of naming trends throughout the inner city. It then examines the attempts orchestrated by the municipal authority Dublin Corporation to change certain street names, not just for political purposes but also as part of a strategy of 'sanitising' certain quarters of the city.

In Part III the emphasis turns towards an analysis of the ways in which a sense of national identity was cultivated in Dublin after 1922. Political developments in the early years of the twentieth century paved the way for the emergence of the Irish Free State, a process that was punctuated by the Easter Rising of 1916, the War of Independence, and the Civil War. As Dublin became capital of the Free State, it played a significant role in the nation-building process. Although the leaders of the first generation after Independence evoked an image of Irish society that was almost exclusively rural, various aspects of the urban landscape did play an important role in marking the transition from the colonial to the post-colonial. What were once the linchpins in both the visual expression of imperial rule and part of a strategy of resistance to the colonial power became instead essential tools in supporting the ideology of the new regime. Chapter 5 considers the powerful symbolism of architecture and urban planning initiatives in giving expression to Dublin's status as capital of an independent country. Particular attention is paid to the debate surrounding the proposed building of a national parliament complex and to the iconographical dimensions of the plans for the reconstruction of the city contained in the work of the Greater Dublin Reconstruction Movement. This chapter also examines a number of other public architectural projects that heralded the arrival of a new epoch in the city's symbolic geography.

The erection of new monuments in the capital after 1922 is discussed in chapter 6. The new administration, just like the previous regime, recognised the powerful role of the past in legitimating authority and forging identity. It was, however, a very different past that was celebrated and commemorated in Dublin after 1922. While old relics of empire were summarily removed, new heroes of the independence struggle were to be erected on plinths throughout the capital and country, and more ancient figures of a heroic Celtic past were unveiled with much ceremony and nationalist triumphalism. Both commemoration and de-commemoration proved integral to the new discourses of power and identity that were mapped out in the post-Independence period. Monuments were not just erected, however, but were also destroyed. Chapter 7, therefore, discusses the processes of de-commemoration whereby monuments of empire were removed by central government or destroyed by dissident republican groups.

The role of street naming and more especially renaming in forging narratives of national identity is discussed in chapter 8. The not always successful attempts made by members of Dublin Corporation to effect such changes is explored, with particular reference to the choice of names preferred for the capital of the Free State. This chapter also places the renaming of Dublin's streets in broader context by briefly examining the naming trends in the city's newly expanding suburbs as well as by engaging with some of the debates which prevailed at national level regarding street and town nomenclature. 1966, the year that marked the triumphant celebrations for the Golden Jubilee of the 1916 Rising along with the bombing of Nelson's Pillar, brings Part III to a close. That year marked a symbolic break after which the iconography of Dublin was to be defined not so much in relation to the struggle that the country had come through five decades earlier, but increasingly by an outward-looking identification with Europe and the outside world.

In the final chapter which draws this work to a conclusion the focus rests upon the contemporary iconography of Dublin some 80 years after the achievement of political independence. It is argued that a shift has taken place from an intensely political iconography to one that is increasingly apolitical and which embodies the changing nature of Irish politics, culture and society at the turn of the twenty-first century. This assertion is explored through an examination of the contemporary iconography of Dublin where public art rather than political sculpture has come to occupy prominent positions in the cultural landscape. Although issues of collective memory, cultural heritage and commemorative practices feature prominently and continue to court controversy, it would seem that the iconography of the city has fundamentally altered as demonstrated through a close reading of the capital's central thoroughfare, O'Connell Street.

Situating Dublin in historical context

Fort of the Dane,
Garrison of the Saxon,
Augustan capital
Of a Gaelic nation,
Appropriating all
The alien brought,
You give me time for thought.

Louis MacNeice, The closing album 1: Dublin, August–September 1939

Over the course of centuries the fortunes of Dublin city have waxed and waned in tandem with broader political developments. During these formative phases of development the cultural landscape came to act as an emblematic site of power and resistance (figure 1, p. 10). It is important to consider the political context within which the cultural landscape was shaped, for the trajectories of Irish politics – the complicated relationship with London, and Dublin's ambiguous status as a city of empire – were to place significant demands on its symbolic fabric. Although its origins stretch back to prehistoric times, the story of Dublin's development as a city begins with the arrival of the Vikings in AD 841.

Since then it has evolved in a number of phases.[1] From the Danish fort created by the Vikings in the ninth century, the city was subsequently shaped by the Anglo-Normans to become a 'Garrison of the Saxon', and from 1171 on the city became a citadel and key base of England's fluctuating power in Ireland. After a time of stagnation and decline in the post-medieval period, the eighteenth century witnessed the emergence of the city as an 'Augustan capital of a Gaelic nation'.

In a century that was to become known as Dublin's 'Golden Age', the urban landscape underwent large-scale redevelopment, the legacy of which remains dominant in the city today.[2] The five Georgian squares and terraces were laid out and named after wealthy landowners or English lord-lieutenants, who were among the major power brokers in the eighteenth-century city and a critical impetus in its development during this period. Important public buildings were commissioned and the axis of the city shifted eastwards away from the old medieval core in the west. It was drawn there by, among other factors, the erection of a new Custom House in the 1780s. Dublin's status as a major port and focus of trade and industry ensured that the city enjoyed decades of prosperity.[3] This is evidenced in the fact that the population increased from an estimated 40,000 people in 1660, to 180,000 in 1800. The Head of State was the British monarch, who was represented in the country by his viceroy, although the real power rested with the chief secretary for Ireland, upon whose advice the viceroy acted. From his offices in Dublin Castle, the chief secretary supervised the administration of a country that was politically and culturally deeply divided and a city whose population in the early eighteenth century was largely Protestant and English speaking, ethnically different from the rest of the country.[4]

Although the Irish legislature became independent in 1782, Westminster retained executive powers and continued to exercise a measure of control over the Dublin parliament. During this period authority in Ireland was directed from London via Dublin Castle, while economic expropriation by planters came in various waves. Ireland was in effect a colony of the British Empire, but was characterised by a form of colonial nationalism espoused by a largely Protestant landowning class that seemed on the verge of defining its own version of Ireland, which was not necessarily British. As Connolly writes, 'it was almost exclusively from the Protestant middle and upper classes of the eighteenth century that a claim to Irish political autonomy was first systematically articulated. In doing so they developed political models, rhetoric and imagery that were to continue – right up to 1914 – to shape the aspirations of the great

majority of Irish nationalists.'[5] The Act of Union of 1800, a legislative response
to the rebellion of the United Irishmen in 1798, served to yoke the two countries
together more closely under the parliament in London and represented the
further integration of Ireland into English political life.[6] Amid these political
developments Dublin became a 'deposed' capital.[7] It fell far behind the rest of
the United Kingdom in terms of industrial and economic development, eventually suffering the ignominy of being overtaken by Belfast as the industrial
capital of Ireland. As the wealthy landowners of the 'Golden Age' deserted the
city in search of better fortunes in London and elsewhere, they left behind them
a city on the brink of a century of poverty, stunted growth and slum dwellings.[8]

After 1800 Ireland effectively became an integrated periphery of the Imperial
State. In many ways, however, it retained colonial status. The viceroy and chief
secretary continued to occupy key positions in the Dublin Castle administration
and worked with a network of departments in Whitehall. Successive political
discontinuities and tensions ultimately ensured that Ireland's status as a colony
was ambivalent and significantly different from, for example, India and
Calcutta, its counterpart capital and Viceregal seat of the British Indian
Empire.[9] A mid-latitude colony of settlement rather than a tropical colony of
exploitation, the settler population in Ireland was small and the colonial
experience was different from that experienced elsewhere and especially in the
non-western world.[10] As Fitzpatrick argues, 'both in form and in practice, the
government of Ireland was a bizarre blend of "metropolitan" and "colonial"
elements. Ireland could therefore be pictured either as a partner in Britain's
empire or as her colony, interpretations that fostered conflicting revolutionary
programmes. While loyalists mustered to defend their metropolitan privileges,
republicans inveighed against their colonial oppressors.'[11]

By 1900 Ireland remained a part of perhaps the greatest empire the world
had ever seen and it seemed clear that whatever change the twentieth century
might bring to the country, it would take place within the context of the
'steady and sterling progress' of a British empire that already ruled nearly a
quarter of the world's population.[12] Dublin was the deposed capital of the
Kingdom of Ireland within the confines of the United Kingdom of Great
Britain and Ireland, and the canvas upon which the British administration and
loyal agents set out to paint a picture of union and loyalty. On the other hand,
it was a city under the local governance of the strongly nationalist Dublin
Corporation which attempted to assert a tangible sense of Irish national
identity upon the urban landscape. Unlike Britain's transoceanic colonies,
Ireland enjoyed a substantial measure of representation in the parliament at

Westminster and was in the eyes of many integral to the composition of the United Kingdom. As Stephen Howe suggests of turn of the century Ireland, it was 'a sphere of ambiguity, tension, transition, hybridity, between "national" and "imperial" spheres', in a manner that was by no means unique but was echoed in places like Canada or Cardiff, Auckland or Aberdeen.[13]

In the early years of the twentieth century the struggle to achieve independence from Britain intensified. While Home Rule was about to be granted in 1914, the onset of the First World War meant that its operation was suspended until after the war.[14] The Easter Rising that took place in April 1916 galvanised a section of the Irish population into revolutionary action, which, after a bloody War of Independence culminated in the creation of the Irish Free State in January 1922.[15] As it stood on the verge of independence the symbolic fabric of the capital exposed the fractured nature of political life in the country, embodied the multifarious nature of Irish nationalism and made tangible the power struggle that persisted between Britain and one of its kingdoms, Ireland.

PART I

*Unravelling the cultural
geographies of landscape*

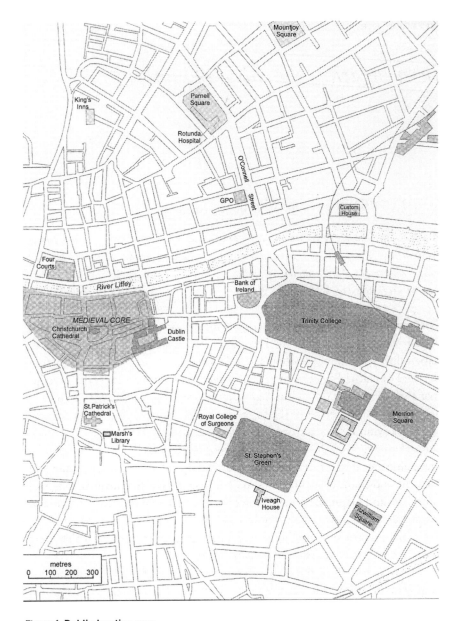

Figure 1 **Dublin location map**

CHAPTER ONE

Interrogating the cultural landscape

Landscape and cultural geography

A landscape is a cultural image, a pictorial way of representing, structuring or symbolising surroundings. This is not to say that landscapes are immaterial. They may be represented in a variety of materials and on many surfaces – in paint on canvas, in writing on paper, in earth, stone, water and vegetation on the ground.[1]

Geographers have long been concerned with the study of landscape in all of its various forms, from the cultural to the physical, the rural to the urban and have followed different paradigms in order to represent its complexity. Within the field of cultural geography it is possible to trace a path from the descriptive analysis of material cultural artefacts of the Berkeley School of cultural geography which thrived in the inter-war period, through to studies of landscape as a symbolic, social construct that have marked the emergence of 'new' cultural geography in the late 1980s and 1990s. Traditionally, cultural geography maintained an interest in the material artefacts evident in the landscape and took the end-point of its inquiries to lie in accounts of the obvious, tangible, countable and mappable phenomena present to the senses of the geographical researcher.[2] This resulted in 'endless studies of house-types, field patterns, log-cabin construction methods, and place-imagery in music', which were 'antiquarian, particularistic and socially irrelevant'.[3]

In the aftermath of the quantitative revolution that marked human geography in the 1960s, cultural geography re-emerged in a radically transformed guise, a product of the cultural turn across the social sciences and the humanities. The seeds of this transformation can be traced to the geographer's increased dialogue with humanism in an attempt to counter and contest the empirical and positivistic nature of the discipline as it then stood.[4] This dialogue was especially important in questioning the traditional conceptions of culture and landscape. Out of the intellectual ferment that followed, a diverse range of new cultural geographies emerged that share a common critical focus. Cultural politics, ideology, hegemony and resistance have come to preoccupy geographers and have opened up new areas of research in subjects as diverse as music, sport, the visual and non-visual arts, performativity and embodiment, as well as cultures of nature and geographies of consumption.[5] At its core the new cultural geography has been marked by a reconceptualisation of the culture concept and novel theorisations of the cultural landscape ensuring that today the sub-field concentrates upon the ways in which space, place and the environment participate in unfolding dialogues of meaning.[6]

For Carl Sauer and his students at the Berkeley School, culture was approached as a 'way of life', an independent and inherited force moulding people and landscape in its image. As he wrote in 1925, 'Culture is the agent, the natural area is the medium, the cultural landscape is the result.'[7] This view led geographers to conceive of landscapes as rather inert imprints or 'containers of culture', and spawned decades of study mainly into rural landscapes which were read for clues of sequent occupancy.[8] Out of the critique of Sauer's super-organic conception of culture, however, has emerged a new understanding of culture as a dynamic process in which people are actively engaged. It has come to be interpreted as a mix of symbols, beliefs, languages and practices that people create, rather than a fixed thing or entity governing humans.[9] It is now argued that the key agents in constructing social life and landscape are not inherited customs, but rather individuals, who 'grasp, interpret and *re-present* their worlds with the use of symbols and vocabularies through which they construct cultures and geographies'.[10] Culture, therefore, encompasses 'ideas, attitudes, languages, practices, institutions and structures of power, and a whole range of cultural practices: artistic forms, texts, canons, architecture, mass-produced commodities and so forth'.[11] Amid this changing context the highly political nature of culture, and the meanings and values attached to it has been underscored.[12]

The cultural landscape has also been reconceptualised and the thrust of the new cultural geography has been to show how landscape forms an integral, dynamic part of social, cultural and political systems.[13] Much more than an areal container of culture, landscape is now conceived as being actively shaped and reshaped, created and destroyed, by people. It acts as a social and cultural production which both represents and is constitutive of past, present and future political ideologies and power relationships.[14] A more interpretative approach to its study has evolved, characterised by a marked sensitivity to its symbolic elements and to their role in the construction, mobilisation, and representation of national and other forms of identity. As Kong observes, landscapes 'are ideological in that they can be used to endorse, legitimise, and/or challenge social and political control'.[15] Much more than a transparent window through which reality may be unproblematically viewed, the cultural landscape is a concept of high tension owing to the political meaning attached to it.[16] It remains the geographer's task to interrogate and interpret this meaning even if it lies buried beneath layers of ideological sediment.[17]

Landscape as text and symbol

As the research lens of the geographer has come to focus on the symbolic power of the cultural landscape, a range of metaphors have come to prominence for understanding the links between landscape and ideology, chief among them the notion of landscape as text and the iconography of landscape.[18] The conceptualisation of landscape as a text draws on literary criticism's traditional point of departure, the triad of author–text–literary critic. When couched in geographical terms by replacing author with agent, text with space or land-scape and the literary critic with the geographer, a useful interpretative approach to landscape is set up. Just as a book is written by an author and is in turn subject to the critique of the literary critic, similarly the landscape or space is 'written' by a set of agents and is subject to the critique of the geographer. The traditional subjects of geographical analysis such as maps, landscapes, and components of landscapes can be viewed as texts, and as products of a wide range of social relations, which are in turn reproduced in the landscape.

The metaphor of 'landscape as text' highlights the authored nature of the world and the fact that cities are 'written' by many agents of power. Just as contemporary literary theory has come to view texts as the product of a 'matrix of social powers', one of the geographer's chief areas of study – landscape – can also be seen as the product of social, political and institutional processes. Agents can be the committee established to erect a statue, the national government that sanctions the opening of a garden of remembrance or the city corporation that votes to change a street name. The analogue meanwhile to the text's critic is the geographer whose task it is to read the city, just as the literary critics' profession is focused on the written text.[19] By making use of the literary triad, geographers can examine the text-like qualities of landscape and see them as transformations of ideology(ies).[20] The cultural landscape may be read interpre-tatively as a text which expresses a distinctive culture of ideas and practices, of often oppositional social groups and political relationships, in order to reveal the ideas, practices, interests and contexts of the society that produced it.[21]

If the urban landscape serves 'as a vast repository out of which symbols of order and social relationships, that is ideology, can be interpreted by those who know the language of the built form', then metaphors such as 'landscape as text' and the 'iconography of landscape' are useful in seeking to excavate their cultural meaning. The iconographical approach to the study of landscape also enables the geographer to see human landscapes as both shaped by and a product of broader social and cultural processes.[22] Defined by Barnes and Gregory as 'the symbolic analysis of visual images that takes into account the cultural context of their production in time and space', the art historian's iconographic method can be applied not only to the painted image but also to the built landscapes and the symbols of which they are composed.[23] As Cosgrove suggests, 'All landscapes are symbolic . . . reproducing cultural norms and establishing the values of dominant groups across all of a society.'[24] The iconographic method explores these meanings by re-immersing landscapes into their social and historical contexts and when carried out successfully it demonstrates the explicitly political nature of representation.

Representing power, re-presenting the past

The idea of the landscape as text and symbol provides an approach to the study of landscape as a series of iconographic discourses embedded in broader cultural and political debates.[25] As an emblematic site of representation, the cultural landscape, whether material or representational, plays a key role in representing the discourses of colonialism, imperialism and forms of nationalism. It also acts as a key player in the heritage process whereby aspects of the past are mapped, often in a highly selective manner, onto the present for a variety of social, cultural and political purposes. In such contexts the symbolic activation of time and space through the public landscape becomes vital to the invention and preservation of a shared collective heritage which is in turn paramount for the legitimation of a national community and state.[26] As powerful regimes and ruling authorities seek to underpin and legitimate their authority, the past and public memory play a crucial role and find tangible representation in the cultural landscape.[27] In order to construct a populace of citizens loyal to and identifying with the state, 'state-builders need to create a homogeneous national culture through the invention of a common history; the introduction of official languages; the creation of national symbols such as currencies, stamps, flags and anthems; the staging of ceremonies and rituals; and the deployment of a wide variety of other cultural forms such as monuments, music and paintings'.[28] The sense of a shared heritage of glorious triumphs and of common suffering is essential in order to maintain communal solidarity and solidify national identity, while at the same time the significance of forgetting certain events is not to be underestimated.[29] Political regimes make effective use of history in order to legitimate and consolidate their dominance. As Azaryahu observes, 'national history is a prime constituent in national identity, while a sense of a shared past is crucial for the cultural viability and social cohesiveness of both ethnic communities and nation-states'.[30]

The past therefore constitutes a mass of cultural capital which can be drawn upon in the context of nation building and empire formation.[31] It represents a cultural and political resource to be seized upon by agents of power in order to shape place identities, support political ideologies and cultivate what Anderson refers to as the 'imagined community' that is the nation.[32] After all, group identity, which is often defined in contradistinction to another group, is both sustained and legitimated with reference to a resource base of shared memories.[33] As Gillis argues, 'identities and memories are highly selective . . . serving particular interests and ideological positions. Just as memory and identity support one another, they also sustain certain subjective positions, social boundaries, and, of course, power.'[34] It is perhaps inevitable, therefore, that aspects of the past should be distilled into icons of identity that are rooted in the cultural landscape and which highlight the historical trajectory of cultural groups in a process that reinforces narratives of group identity.[35] The dynamic relationship between history and geography is demonstrated when national monuments, public buildings and streets celebrating national heritage are inserted into the landscape in a manner that maps history onto territory. These landscape elements demonstrate the ways in which representations of place are intimately related to the creation and reinforcement of official constructions of identity and power and to the whole question of empowerment.[36] Within official public landscapes, state-sanctioned monuments, architectural and urban design initiatives along with naming strategies act in different ways and to varying extents as spatialisations of memory, making tangible specific narratives of nationhood and, as Bell points out, reduce fluid histories into sanitised, concretised myths that anchor the projection of national identity onto physical territory.[37]

Contested landscapes: geographies of resistance

While dominant regimes use history, heritage, memory and various landscape elements in order to memorialise specific narratives of nationhood, such narratives can also be challenged and resisted by those less powerful, especially in colonial contexts. Consequently, in contested political contexts where some groups do not identify with the dominant elite and seek empowerment, the cultural landscape often becomes a site of conflict and contestation.[38] Although traditionally geographers have tended to focus upon the historical geography of colonialism with the emphasis firmly fixed on the processes and patterns of colonial domination, this has been destabilised in recent years by an attempt to accommodate what Yeoh refers to as a 'politics of space in the colonised world where people resisted, responded to and were affected by colonisation'.[39] Hence, geographers are now more sensitive to the complexities and multiplicity of resistances that prevail at the contact zone between the coloniser and the colonised. Within cultural geography there has been a repositioning from an engagement purely with geography and empire where the emphasis is firmly focused on the colonising power, to a consideration of the multiple historical geographies of the colonised world.[40] Thus, landscape is now read as a site of conflict and collision, negotiation and dialogue between the coloniser and the colonised. Moreover, those symbols of dominance, which make concrete the power of prevailing regimes, also serve as focal points around which the marginalised and less powerful can orchestrate opposition and resistance.

In post-colonial and post-independence contexts, the re-evaluation of heritage that invariably prevails ensures that attempts are often made to unravel the iconography of the previous regime and to re-inscribe new narratives of nationhood. Newly independent states often share in common a desire to reconstruct and destroy the inherited legacies and official iconographic landscapes that defined national heritage during previous regimes.[41] Consequently, rituals of revolution become immensely significant in creating new symbolic spaces of power and making tangible radical political shifts. As Lefebvre observes, 'A revolution that does not produce a new space has not realised its full potential; indeed, it has failed in that it has not changed itself, but has merely changed ideological superstructures, institutions and political apparatuses. A social transformation, to be truly revolutionary in character, must manifest a creative capacity in its effects on daily life, on language and on space.'[42] Agents of power in newly independent states are acutely aware of the symbolic impact of particular landscape elements when married with the legacy of the pre-colonial past in affirming ideology and reinforcing nationhood.

New symbols of legitimacy are sought out and narratives of nationhood become embedded in the cultural landscape. The erection of monuments, the destruction of earlier erected statues redolent of the previous regime, the naming and renaming of streets, as well as the building of new public buildings, and in some cases the building of new capital cities, all become important in cultivating national identity and promoting national pride. It is to the role of such 'icons of identity' in forging symbolic landscapes and shaping narratives of identity that this chapter now turns.

Interrogating landscape's icons of identity

Symbols are what unite and divide people. Symbols give us our identity, our self-image, our way of explaining ourselves to others. Symbols in turn determine the kinds of stories we tell; and the stories we tell determine the kind of history we make and remake.

Mary Robinson, Inauguration speech as President of Ireland, 3 December 1990

As an emblematic, socially constructed site of representation, the cultural landscape and the signs and symbols of which it is composed play a crucial role in legitimating particular political and social orders and in contributing to narratives of group identity. In this section particular attention is directed towards four aspects of the cultural landscape: public statuary, street names, architecture and urban planning initiatives. Where cities evolve in contentious political circumstances and make the transition from the colonial to the post-colonial, these aspects of landscape take on particular significance. In different ways and to varying extents they play a crucial role in shaping the iconography of landscape and act as significant sources for unravelling the geographies of political and cultural identity. These 'icons of identity' also draw to differing extents upon the cultural capital of the past in order to reinforce the dominance of particular ruling authorities, while at the same time they can act as focal points around which resistance and opposition can be channelled, especially in the context of post-independent cities when they become implicated in strategies of nation building.

Public statuary and the politics of commemoration

Among the most strikingly symbolic features of any town, city or urban landscape are the public monuments which line its streets and dot its squares. Commemorating an individual or an historic event, public monuments are not merely ornamental features of the urban landscape but rather highly symbolic signifiers that confer meaning on the city and transform neutral places into ideologically charged sites. While the ancient Greeks used them as a means of conferring honour on esteemed members of society, it was not until the mid-nineteenth century that public statues took on particular significance as a means of celebrating a nation's past. Until the outbreak of the First World War, statues served as a symbolic device of enormous popularity, and imposed the ideals and aspirations that they represented onto the public consciousness in a manner that other cultural signifiers could not.[43] The intense nationalism of these years gave rise to widespread and sustained attempts across Europe and North America to commemorate national histories through monuments as 'statues sprouted up on the public thoroughfares of London at a rate of one every four months during Victoria's reign. The streets and public squares of even modest-sized German towns bristled with patriotic sculpture: in a single decade some five hundred memorial towers were raised to Bismarck alone.'[44]

The frenzy of monument building that occurred in Europe stemmed largely from the fact that governments recognised their key role as foci for collective participation in the politics and public life of villages, towns and cities. Statues served to strengthen support for established regimes, instilled a sense of political unity and cultivated national identity. Agulhon uses the term 'statumania' to describe the frenzy of monument building that occurred in France for example, where in Paris city officials considered imposing a moratorium on the erection of new monuments.[45] Sculpture spread into the arena of public secular space as European governments recognised that public monuments acted as the foci for collective participation in the politics and public life of villages, towns and cities. Moreover, monuments strengthened popular support for established regimes, instilled a sense of political unity and cultivated a sense of national identity.[46] Lerner captures the symbolic essence when he notes that:

> embedded within the monument is a particular way of staging politics that is centred on the spectacle or visual display. With its emphasis on representing human forms, the monument reveals two important terrains upon which political power and the form of the nation rest: the spectacle of politics and the

public display of the body. Like the bronze and stone figures themselves, these terrains have tremendous resilience over time.[47]

For those concerned with understanding the dynamics at work in shaping the historical and the contemporary urban landscape, acts of memorialisation and the statues which give tangible expression to them are of much significance. The objects of a people's national pilgrimage, monuments are signifiers of memory which commemorate events or individuals but also 'mark deeper, more enduring claims upon a national past as part of the present . . . monuments may become both historical symbols of nationhood and fixed points in our contemporary landscapes'.[48] They effectively map history onto territory, creating in the process a repository of common memories in a manner that is integral to the ideological rhetoric of nationalism.[49] As Sandercock suggests, we erect sculptures to dead leaders, war heroes and revolutionaries because 'memory, both individual and collective, is deeply important to us. It locates us as part of something bigger than our individual existences, perhaps makes us seem less insignificant . . . Memory locates us, as part of a family history, as part of a tribe or community, as a part of city-building and nation-making.'[50] If the city is a repository of collective memory, then public statues make an important contribution to its memory bank while focusing attention on specific places and events in highly condensed, fixed and tangible sites.

Dynamic sites of meaning and memory which transform neutral spaces into sites of ideology, public statues not only help to legitimate structures of authority and dominance but are also used to challenge and resist such structures and to cultivate alternative narratives of identity. Just as public statues served throughout the late nineteenth and early twentieth centuries as a means of cultivating popular support, making power concrete in the land-scape, the medium was also employed, paradoxically, by groups at odds with established regimes as a means of challenging the legitimacy of governments and objectifying the ideals of revolutionary movements. For the very qualities that make public statues so valuable in building popular support for established regimes also make them a useful target for those who wish to over-throw them.[51] This is particularly the case in post-colonial countries emerging from beneath the shadow of the colonial enterprise or ideological domination, the urban landscapes of cities in Eastern Europe being a case in point. In the wake of the fall of communism the dissolution of the 'Iron Curtain' precipitated the mass removal of a vast number of public statues dedicated to the mono-lithic figures of communist rule such as Marx, Stalin and Lenin. These statues, when in place on the streets and major thoroughfares of Eastern Europe, exercised considerable propagandist strength in their choice of style, scale and position. Their consequent removal from the urban landscape delivered a cogent symbolic message.[52] Figure 1. 1 illustrates the 'life cycle' of a public monument and provides a useful framework around which to base an interpretation of its meaning.

Figure 1. 1 **The life cycle of a public monument**

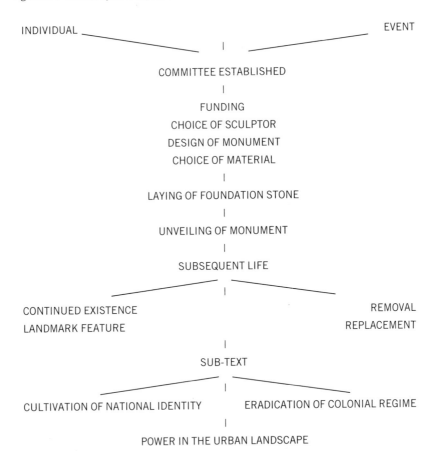

Street naming and the politicisation of space

Another important means through which societies, both today and in history, have utilised the past as a political, cultural and social resource is through the naming of towns, cities and, in particular, streets. Affixing names to places is inextricably linked with nation building and state formation and consequently sweeping changes in the naming process reflect ideological upheavals.[53] Turbulent political events of twentieth-century Eastern Europe once again provide a ready example. Formerly St Petersburg in the days of Tsarist Russia, Leningrad was renamed under communism. In a gesture of much symbolism, however, the city reverted back to St Petersburg in a referendum in June 1991. In both colonial and post-colonial contexts the names given to streets take on great symbolic significance in tandem with the obvious functional significance of a street name. Just as public statues are central to the creation of national mythologies and the invention of tradition, so too street names, albeit in a different, less audacious way, manage to commemorate past events and figures and are used by political regimes in order to legitimate and consolidate their dominance and reinforce their authority. Traditionally street names tended to be vernacular in origin, designating geographical orientation and urban function and often drawing upon facts of local topography and history. This began to change in the eighteenth century, however, when, with the introduction of the postal service, the responsibility for naming fell into the hands of official-dom. Gradually street names began to take on greater political significance, a trend which gathered pace in the latter half of the nineteenth century, when naming became a prominent feature of the age of modern nationalism, colonisation and empire building.

As dominant powers set about cultivating a sense of collective identity and creating a shared past, official and authorised versions of history were often made concrete in the urban cultural landscape though the use of street names. Naming streets after famous persons or events perpetuated in the streetscape and in the minds of inhabitants the memory of historical figures or events deemed to be worthy of remembering by those in charge. In effect, 'commemorative street names, like their alphanumerical counterparts, provide locational infor-mation, but they also have the function of perpetuating, reifying and constructing a view of the past'.[54] The selection of street names, therefore, is an inherently political procedure determined by ideological needs and political power relations and always implemented by agents of ruling power. Street names, as Azaryahu suggests, conflate history and geography and bring the past that they comme-morate into the everyday language and ordinary settings of human life, thereby transforming history into a feature of the natural order of things.[55] Naming plays a crucial role in processes of empire building, as well as in the ritual of revolution and rebellion that marks the emergence of post-colonial nations when renaming and divesting the namescape of past historical associations becomes an essential tool. Any discussion of naming, therefore, must also consider the phenomenon of renaming and the role of street names in challenging the legitimacy of historical traditions and in building national identity.[56] Street names can serve as focal points around which expressions of dissent and opposition can be anchored. Citizens can demonstrate their opposition to an established regime and directly challenge authority by resisting the names imposed by a higher authority, while renaming streets, just like destroying monuments, has an immediate effect on the politics of landscape and serves as an act of political propaganda of enormous symbolic value, ensuring that citizens are speedily confronted with new political realities.

In essence the naming process, 'embodies some of the social struggle for control over the means of symbolic production within the built environment'.[57] Names effectively turn spaces into places, and reflect the operation of power. They 'give colour, meaning, symbolism to the world around us'.[58] This applies as much to the suburban housing estates of the late twentieth century as it does to the names that are assigned to streets in the heady days after the achievement of independence. Not only signs *to* the city, street names are also very much signs *of* the city, which like all the other expressions of urban civilisation supply symbolic capital that cities spend in many different ways. As Entrikin observes, 'in both everyday conversation and in literary narratives, place names have a semantic depth that extends beyond the concern with simple reference to location or to a single image.'[59] Ultimately, the spatial distribution of names and the individuals or events that they commemorate, when set within the context of related aspects of the built environment, serve as sensitive indicators of the links between politics and the landscape. Debates and issues of contestation over naming can be seen as symbolic representations of larger power struggles between competing interests and identities. In their ability to transmit meaning, therefore, street names are integral to the iconography of landscape.

Representing power: architecture and urban design in the city

Just as street naming and public statuary are crucial to the affirmation of particular ideologies in different political contexts and to the construction of narratives of national identity, so too architectural and urban design initiatives play an important role. The politically motivated construction and reconstruction of urban spaces have long fascinated geographers interested in exploring the interrelations between ideology and landscape. Many urban landscapes have been laid out, if not at their very beginning then at some point in their existence, according to a plan. This plan is a framework around which the city is destined to evolve. It plots, for example, the layout of streets and squares, of prominent buildings and monuments, and it makes provisions for housing, for public parks and for private business space. More than all of these things, however, the plan of a city is bound up with the politics of power and identity. Often accompanied by a written text, the city plan is also underpinned by a sub-text. As Kenny argues, 'a planning document, possibly more than any other written text, articulates the ideology of dominant groups in the production of the built environment'.[60]

For centuries the city form has been planned and manipulated in order to represent power and ideology. Features such as squares and road patterns, whether axial, orthogonal or gridiron, contribute to a geometry that is radically different from the curves and undulations of the natural landscape but which represent human reason and the *power* of intellect.[61] Countless examples of civic design initiatives the world over demonstrate the fact that the city plan, the skeleton around which an urban landscape develops, is an orchestrated piece of work with much symbolic underpinning. This includes both the plans that made the transition from the drawing board to reality, and also those that remained ideas. Cities such as Washington DC, New Delhi or Canberra, the European cities like Berlin, Moscow and Rome that were reshaped during the totalitarian regimes of the 1930s or cities of North Africa shaped during colonial rule, each embody in their planned layout a considerable measure of symbolic potency.[62]

Within the structures of the city plan, the specific significance of architecture demands attention. More than merely assemblages of bricks and mortar, buildings are invested with meaning and possess a particular representational and aesthetic purpose alongside their inherently practical function. As Jencks suggests of architecture, 'not only does it express the values (and land values) of a society, but also its ideologies, hopes, fears, religion, social structure, metaphysics'.[63] Buildings effectively communicate by symbol the culture of which they are a part, representing that culture as well as housing it in a manner that draws upon the symbolic capital inherent in the architectural form.[64] A policy document issued by the Irish government in the mid-1990s suggested that the architecture of a people is 'an expression of its culture, and an integral part of its identity, as well as being a response to the requirement for shelter. At its highest level it takes its place among the arts as an expression of the human spirit.'[65] It is, therefore, one of the tasks of the cultural geographer in reading cities to uncover the meaning and politics invested in the buildings which prevail in the cultural landscape.[66] In interpreting such buildings it is necessary to situate them amid the cultural and political context in which they were built. Location, architectural style, function, design, ornamentation and iconography, the type of building material employed and the ceremony and ritual associated with their construction and opening, as well as how their function changes over the course of time, are all aspects which become important when reading the built environment.[67] Public buildings provide an important means of gaining access to the meaning embedded in the urban landscape. As Mumford observed, 'In the state of a building at any period one may discover, in legible script, the complicated processes and changes that are taking place within civilisation itself.'[68]

Reading icons of identity in broader context

The symbolic approach to understanding landscape underscores the fact that it is a highly complex discourse in which a whole range of economic, political, social and cultural issues are encoded and negotiated.[69] As Seymour argues, 'the view comes from somewhere, and both the organisation of landscapes on the ground, and in their representations, are and have been often tied to particular relationships of power between people'.[70] The powerful role of public monuments, street naming and architectural and urban design initiatives in shaping particular narratives of identity across a variety of spatial scales and in an equally broad range of political, social and cultural contexts has been the subject of a burgeoning body of research in historical and cultural geography.[71] This work demonstrates that signs and symbols are integral in giving concrete representation to forces of dominance as well as of resistance and opposition. Although central to processes of empire formation and nation building, landscape represents more than the impress of state power or elite ideologies. Rather, the cultural landscape embodies *many* interwoven layers of power and overlaps with issues of race, gender, class and local identity politics. Cultural and historical geographical research on the symbolic meanings of landscapes has therefore attempted to disclose the dynamic nature of the relationship between landscape, memory and power in the construction of nation identity. It has shown that landscapes, both representational and material act as a means of articulating discourses and identities and as a medium for the representation of competing discourses among different social and political groups.

On the power of street naming and renaming, for example, Yeoh has demonstrated that affixing names to places in colonial and post-colonial Singapore was inextricably linked to nation building and state formation.[72] One hundred and fifty years of colonial rule in Singapore conferred on its landscape an official network of street and place names which reflected the mental images and ideological purposes of the dominant culture.[73] Street names recorded the colonial imagination and commemorated the British monarchy, European settlers, Governors and public servants as well as deserving citizens who were deemed to have contributed to the urban development of the city under colonial rule. They honoured the priorities of powerful European namers rather than those of the people living in the places so named.[74] After independence in 1965 Singapore emerged as a complex and multi-racial community with little sense of a common history; naming became central to the nation-building project. The vision of the new state therefore was to build a nation based upon principles of multiracialism, multilingualism and multi-religiosity, which would accommodate the Chinese, Malay, Indian and other races. In post-colonial Singapore, the multi-cultural nature of society meant that in terms of nation building it was not just a matter of stamping out the colonial imprint and replacing it with a new one. Instead, the authorities had to take close account of the multiracial and multi-lingual nature of society.

In colonial and post-colonial Zanzibar, the politics of naming also proved highly contentious. Myers has demonstrated this in his work on the politics of naming and renaming in the Ng'ambo district, an area of the city created during the British colonial period in the early twentieth century. During this period the naming of the Ng'ambo urban landscape played a critical role in British efforts to secure legitimacy, while in post-colonial times names were used to express the new socialist order that came to prevail. [75] Similarly, Cohen and Kliot's work illustrates the way that the Israeli nation state selected place names for the administered territories of the Golan, Gaza, and West Bank in order to reinforce national Zionist ideologies. Between 1977 and 1992 Biblical and Talmudic place names were introduced by the ruling right-wing Likud to project Israel as the rightful heir to the Holy Land.[76] In Europe, Azaryahu has shown that the naming and renaming of streets in Berlin contributed to the building of German nationhood following unification in the nineteenth century, especially through the commemoration of historical and cultural heroes. He demonstrates that spatial commemoration through nomenclature reflects power relations, while also serving to anchor popular resistance. For example, 'the use of an alternative appellation can be an act of revolt; in the city of Lodz, Poland, residents rejected one Communist name, Street of the Siege of Stalingrad, preferring its former title, Street of November 11'.[77] Shifting time periods to the late twentieth century, Azaryahu has also explored the process of renaming in East Berlin following the reunification of Germany in 1990.[78] He suggests that the renaming of streets during 1991 'represented the effort of the district politicians to rid the cityscape of salient manifestations of the Stalinist past. This mainly meant the eradication of appellations that directly pertained to Soviet political hegemony and to the political history of the GDR.'[79] However, not all street names redolent of the previous regime were changed and the renaming strategy did not attempt to obliterate the revolutionary heritage of German socialism completely from the streetscape.

The inter-linkages between class politics and naming strategies are teased out in Pred's analysis of official and popular street names in nineteenth-century Stockholm. This work uncovers working-class resistance to the naming policy of the upper and middle classes.[80] He notes that in the 1890s, official street name revisions 'were heavily laden with the ideology shared by elements of the financially powerful and well-to-do middle classes: "patriotic and historical names"; "Nordic mythology" . . . "famous Swedish authors" and "prominent men of technology and engineering"'.[81] This dominant ideology was often resisted by the working class through a variety of means including the employment of street names used before the official renaming, the colloquial use of names that denigrated the upper classes and political battles to fend off the renaming of streets. Berg and Kearns, meanwhile, have demonstrated that the process of naming places involves a contested identity politics of people and place that is bound up with issues of race and gender.[82] They argue that the names applied to places in New Zealand reinforced claims of national ownership, state power and masculine control and legitimated the dominance of a hegemonic Pakeha masculinism.[83]

Given the upsurge of interest in the production of landscape and its relationship with concepts of memory, heritage and identity, it is perhaps not surprising that public statues should come under the geographical microscope. Pioneering papers by Johnson, Heffernan, Withers and more recently Peet, Osborne and Cosgrove and Atkinson have illustrated that monuments are bound up with the metaphor of landscape as a way of seeing. They are in effect one way of seeing or of reading into the power politics at work in the creation of the landscape. This research has built on a tradition established by historians, ethnographers and anthropologists and has contributed directly to the reju- venation of cultural geography. Their work highlights the multifaceted nature of the seemingly innocuous public monument, from serving as sites of local protest and contestation,[84] to constituent elements of a national war.[85] Graham's work has highlighted that even within the Protestant community of Ulster expression of identity in the cultural landscape was realised differently,[86] while Johnson has also explored the ways in which national imagined communities are constructed in public statues.[87] In his seminal work on the Sri Lankan city of Kandy, Duncan has shown that the symbolic decolonisation of the city after Independence was articulated through the removal of statues dedicated to British political figures which were then replaced with statues in honour of important Sri Lankan nationalist leaders.[88]

Other studies have focused more particularly on specific monuments, but have been no less illuminating. Atkinson and Cosgrove's analysis of the Vittorio Emanuele II monument in Rome demonstrates the role of this monument as a vehicle for the official rhetoric of a united and imperial Italy.[89] Osborne's account of the George Étienne Cartier monument in Montreal illustrates the means by which monuments serve as devices that ensure particular histories and geographies become encoded in the landscape.[90] Similarly, Peet has pointed to the meaning vested in the public monument in his exploration of the Daniel Shays memorial in Petersham, Massachusetts.[91] All of these studies contribute to a growing literature on public statues within cultural geography. Moreover they articulate key aspects of the power of the public monument from the local scale through to the expression of national identity.

The role of public buildings and urban planning in contributing to the iconography of landscape has been examined in a range of research papers, especially since Harvey's seminal work on the Basilica de Sacré Coeur in Paris.[92] His reading of the Sacré Coeur highlights the contested political meaning of the site at Montmartre, where both conservatives and communards could lay symbolic claim; it is an early and much quoted example that demonstrates the power associated with the built form. This power emanates not just from the physical presence of the building but also from the fact that it was built on an especially symbolic site where the martyrs of the Paris Commune Uprising of 1871 had been massacred. Events conspired to make the Sacré Coeur 'a monument to the triumph of bourgeois values, the ascendant power of the right-wing clericalism and the installation of sentimental piety in opposition to the secular state'.[93] Domosh, meanwhile, has demonstrated the symbolic meaning associated with public buildings in her work on landscapes of corporate architecture and in particular the skyscraper. She has argued that 'they were invested with meaning by those who viewed them, but they of course were also created by people with certain intentions, intentions that developed out of circumstances particular to late nineteenth century New York'.[94] Her work demonstrates that corporate bodies were engaged in a complex activity of communicating by symbol the culture of which they were a part. This is echoed in a range of other studies, including, for example, Black's work on another form of corporate architecture, bank buildings, and his reading of the symbolic capital of the London and Westminster bank headquarters from 1836–8,[95] and in Woolf's reading of the Paris Opera House in the nineteenth century where she notes that no other building could have been more symbolic of the Second Empire, 'it mirrors Parisian life between 1852 and 1870 . . . a building such as this could be said to capture a moment in time'.[96]

The construction of buildings has traditionally been an important element in nation building and empire formation. As Al Sayyad demonstrates in his book *Forms of Dominance*, buildings have served to help establish national consciousness in both colonial and post-colonial eras.[97] Just as public statues and street naming represent the power and domination of the ruling elite, as well as serving as tools in strategies of resistance, public buildings play an equally significant role. In an account of the politics of space in post-independence Helsinki, Haarni describes the symbolic significance of the buildings erected in Helsinki when he writes of how the soul of the young independent nation is revealed in the psychoanalysis of the Parliament House. He goes on to state that in the inauguration of the building in 1931, the Finnish President, Svinhufvud, said that the Parliament House 'should be the pro-moter of a high spirit of patriotism and noble ideas'.[98] This is echoed in Lewandowski's work on the re-fashioning of the urban landscape of Madras after 1949 when Tamil nationalism was inscribed into the urban landscape through the erection of statues, temples, public buildings and the renaming of streets.[99] Vale meanwhile demonstrates in his wide-ranging study of archi-tecture, power and national identity that parliament buildings in particular serve as symbols of state.[100] Through an exploration of parliamentary complexes in capital cities on six continents he shows how buildings housing national government institutions are products of the political and cultural balance of power within pluralist societies.

The examples presented above suggest that landscape iconographies can be read and interpreted as symbolic representations of larger power struggles. They highlight the fact that landscape plays a crucial role in the process of cultivating a sense of a shared heritage, for it is within the domain of public space that mythical evocations of the past are fleshed out and where grand urban design and architectural initiatives proclaim the onset of new epochs, together with the death of older ones. The iconic aspects of landscape discussed provide a composite framework for interpreting the symbolism of the urban cultural landscape and they demonstrate the ways in which history, heritage and memory can become fused with landscape in order to consolidate and legitimate structures of power and authority. In Part II these ideas are explored in relation to the city of Dublin before political independence.

PART II

City of empire, site of resistance:
the iconography of Dublin before independence

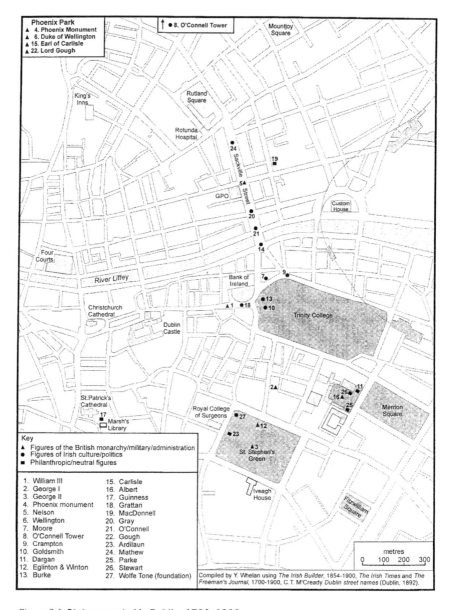

Phoenix Park
▲ 4. Phoenix Monument
▲ 6. Duke of Wellington
▲ 15. Earl of Carlisle
▲ 22. Lord Gough

↑ ● 8. O'Connell Tower

Mountjoy Square

King's Inns

Rutland Square

Rotunda Hospital

● 24

Sackville Street

● 19

5 ▲

GPO

● 20

● 21

● 14

Custom House

Four Courts

River Liffey

Bank of Ireland

7 ● ● 9

● 13

Christchurch Cathedral

▲ 1 ● 18

● 10

Trinity College

Dublin Castle

St. Patrick's Cathedral

17

Marsh's Library

2 ▲

Royal College of Surgeons ● 27

2 ▲

16 ▲ 26 ● 11

25

Merrion Square

▲ 12

● 23

▲ 3

St. Stephen's Green

Iveagh House

Fitzwilliam Square

Key
▲ Figures of the British monarchy/military/administration
● Figures of Irish culture/politics
■ Philanthropic/neutral figures

1. William III
2. George I
3. George II
4. Phoenix monument
5. Nelson
6. Wellington
7. Moore
8. O'Connell Tower
9. Crampton
10. Goldsmith
11. Dargan
12. Eglinton & Winton
13. Burke

15. Carlisle
16. Albert
17. Guinness
18. Grattan
19. MacDonnell
20. Gray
21. O'Connell
22. Gough
23. Ardilaun
24. Mathew
25. Parke
26. Stewart
27. Wolfe Tone (foundation)

metres
0   100   200   300

Compiled by Y. Whelan using *The Irish Builder*, 1854-1900, *The Irish Times* and *The Freeman's Journal*, 1700-1900, C. T. M'Cready *Dublin street names* (Dublin, 1892).

Figure 2.1 **Statues erected in Dublin, 1700–1900**

CHAPTER TWO

Representing Empire: the monumental legacy of monarchy and military

Reading the monumental legacy

Hence nations have wisely sought to place
Records of great men, for their youth to trace,
That all beholding them might bear in mind
Virtues and actions of the nobler kind.
Erin's fair capital is richly blest
With high memorials thus to fire the breast.[1]

At the turn of the twentieth century, when Ireland stood on the cusp of political developments that would eventually lead to independence, citizens of Dublin lived in a city peopled with a variety of figures cast in stone. A close examination of the figures commemorated in the city at this time suggests that these monuments, far from being merely ornamental features of the cityscape, constituted overtly political and highly contentious aspects of the urban fabric. Of those statues mapped in figure 2.1, the earliest were those dedicated to figures of the British monarchy and military establishment and it is these monuments that form the focus of this chapter (table 2.1). In 1701 the first, and for some time the only equestrian statue in the city, was erected in commemoration of King William III. That statue, in the individual that it commemorated, the style that was chosen to represent the king and the choreography that surrounded its unveiling as well as subsequent anniversary celebrations, set a trend for much of the eighteenth and early nineteenth centuries. Successive memorials were unveiled in prominent locations, dedicated to British monarchs and heroes of the British military establishment, namely Lord Nelson and the Duke of Wellington. These statues served as signifiers that expressed the links with the British Empire, both before and after the Act of Union in 1801. Largely a result of the initiative of local, loyal institutions, these monuments constituted the first layer in the city's monumental fabric which was to remain unchallenged until the latter decades of the nineteenth century, when statues dedicated to figures drawn from the contrary sphere of Irish culture, literature and the various strands of nationalist politics found their way onto prominent public thoroughfares in the city.

Table 2.1 **Statues erected in Dublin 1701–1850**

Individual/Sculptor	Date	Location
King William III/Gibbons	1701	College Green
King George I[1]/Van Nost	1722	Essex Bridge
King George II/Van Nost, the younger	1758	St Stephen's Green
Lord Nelson/Kirk and Wilkins	1809	Sackville Street
Duke of Wellington[2]/Smirke	1817	Phoenix Park

1   This statue was moved to the garden of the Mansion House on Dawson Street in 1782.
2   By the time of completion in 1861, a number of other sculptors and artists had become involved.

Memorialising the British monarchy: an eighteenth-century inheritance

How absurd these pompous images look, of defunct majesties, for whom no breathing soul cares a halfpenny! It is not so with the effigy of William III, who has done something to merit a statue. At this minute the Lord Mayor has William's effigy under a canvas and is painting him of a bright green, picked out with yellow – his lordship's own livery.[2]

Although somewhat scathing in his remarks about Dublin's statuary and its 'pompous images', Thackeray in his *Irish sketchbook* touches on one of the salient aspects of the city's urban landscape, its monuments, and more specifically those dedicated to members of the British monarchy. Of the statues in place by the turn of the twentieth century, three had been erected in the early decades of the eighteenth century in commemoration of members of the British royal family, kings William III, George I and George II. Each was unveiled with much pomp and ceremony in a prominent location in the city centre, and served to express Dublin's status as a city of the British Empire and a loyal dominion.

The first to be unveiled was that dedicated to, 'our own King William III, in College Green, with the steed of such wonderful anatomical development'.[3]

King William III, College Green

Here in just right the equestrian statue stands Of Orange's great Prince, deliverer of these lands, Whose name reminds us of those troubles great In which the second James involved our state.[4]

King William III (1650–1702) arrived in Ulster in 1690 and played a key role in securing a Protestant victory over the Catholic King James II at the Battle of the Boyne in July of that year. His intervention ensured that the Protestant population regained control of Dublin.[5] It is perhaps not surprising then to discover that the birthday of King William, 4 November, a date which coincided with his arrival in England, had been a focus of annual celebration in the capital since 1690. At a time when the struggle with the Jacobites was still precariously balanced, 'the Lords Justices authorised a major public celebration on 4 November in honour of King William's fortieth birthday'.[6] Gilbert provides a detailed description of the events that occurred every year on 4 November, when:

> In the morning the English flag was displayed on the tower at Dublin Castle; the guns in the Phoenix Park were fired, answered by volleys from the corps in the barracks, and by a regiment drawn up on College Green; all the bells in the town rang out. At noon the Lord Lieutenant held a levee at the Castle, whence, about 3 p.m., a procession was formed, the streets from the Castle being lined with soldiers. The procession, composed of the Viceroy, Lord Mayor, Sheriffs, Aldermen, Lord Chancellor, Judges, Provost of Trinity College, Commissioners of Revenue, and other civil and military officers, together with those who had been present at the Castle, moved through Dame street and College Green to Stephen's Green, round which they marched, and then returned in the same order to College Green, where they paraded thrice round the statue, over which, after the procession had retired, three volleys of musketry were discharged by the troops.[7]

The erection of the statue in July 1701 was intended to provide a more tangible dimension to the annual affirmations of loyalty. It also introduced a new cele-bratory day to the calendar which was kept thereafter as a public holiday when the memory of King William was celebrated with great gusto in the capital.[8]

In his preface to volume VI of the *Calendar of the ancient records of Dublin* (CARD), Gilbert notes that, 'among the many important matters which during that period [1692–1716] occupied the attention of the municipal council the following may be mentioned. Early in the year 1700 the civic assembly of Dublin resolved to erect a statue of William III "in copper or mixed metal."' It was intended to place the statue on a pedestal in the Old Corn market of Dublin, 'defended by iron banisters'.[9] The Dutch sculptor, Grinling Gibbons, was commissioned to sculpt the form of the King and in July 1700 it was ordered 'by the authority of the said assembly that his majesties statue, when finished, be erected in College Green'.[10] By June 1701, the equestrian statue of the king, which represented him in Roman armour astride his horse, crowned with a laurel wreath and carrying a truncheon, was completed and arrangements were made for the unveiling ceremony.

The city assembly voted to unveil the monument on 1 July 1701, the anniversary of the Battle of the Boyne. On that day, 'the Lord Mayor and Sheriffs were authorised to disburse one hundred pounds in entertaining the Lords Justices, and bestowing some wine to be publicly let run on the said place'.[11] The unveiling was marked by a gathering of all of the members of the city assembly at the Tholsel at 4 p. m. after which there was a formal procession to the College Green site, led by the city musicians and the grenadier companies of the Dublin Militia. After the civic officials had mustered at College Green, the Lords Justices arrived at the statue. The entire assembly then marched around the statue three times. Gilbert captures the sense of the occasion when he writes:

> After the second circuit, the Recorder delivered an eulogy on King William, expressing the attachment of the rulers of Dublin to his person and government; on the conclusion of which oration a volley was fired by the grenadiers . . . At the termination of the third circuit round the statue, the Lords Justices, the Provost and Fellows of Trinity College, with Williamite noblemen and gentry, were conducted by the Lord Mayor, through a file of soldiers, to a large new house on College Green prepared for their reception . . . The surrounding crowds were regaled with cakes thrown amongst them, and several hogsheads of claret were placed on stilts and set running. The Lords Justices, attended by the civic officers, then proceeded to the Lord Mayor's house, where an entertainment was prepared for them, the nobility and ladies; after which fireworks were discharged, and the night concluded with the ringing of bells, illuminations, and bonfires.[12]

Once in place on College Green the statue became a focal point of both the annual birthday celebration of King William on 4 November, as well as the 1 July anniversary of the success at the Battle of the Boyne. These events were celebrated throughout the course of the eighteenth century, and served to express allegiance and loyalty to the empire. The formation of a number of Williamite Associations in the early decades of the eighteenth century, as well as of the Volunteers and the Orange Society in the 1770s, gave the statue renewed ceremonial importance. Their annual musters in the vicinity of the statue commenced on 4 November 1779, 'when all the bells in the city were rung at the opening of the day, and the citizens appeared decorated with orange ribbons'.[13] The event is captured in the painting illustrated in plate 2.1.

Plate 2.1 **King William III, College Green**
Erected in 1701 and sculpted by Grinling Gibbons
From Francis Wheatley, *The Dublin volunteers in College Green, November 4, 1779*, National Gallery of Ireland

While the statue served as a focus of much loyalist celebration, once in place in the public domain it became a focal point for demonstrations of resistance. In the early part of the century Gilbert records that the spirit of Jacobitism which existed in Dublin:

> combined with a love of mischief, and a desire to revenge the insult offered to the college by the King's back being turned towards it, provoked repeated

indignities upon the statue. It was frequently found in the mornings decorated with green boughs, bedaubed with filth, or dressed up with hay; it was also a common practice to set a straw figure astride behind that of the King.[14]

Throughout the remainder of the eighteenth and into the nineteenth century countless examples are recorded of the abuses heaped upon the statue. One of the earliest was the work of three students of the nearby Trinity College, who in 1710 'covered the King's face with mud, and deprived his majesty of his truncheon'.[15] Subsequently, the city council made the unanimous resolution that 'all persons concerned in that barbarous act are guilty of the greatest insolence, baseness, and ingratitude'.[16] They later offered £100 from the government and £50 from the city for the discovery of the iconoclasts.[17] The defacement was discovered to have been the work of three students of Trinity College – Graffon, Vinicome and Harvey. They were expelled from the university and while Harvey managed to escape, the other two confessed their guilt and were condemned to six months' imprisonment, a fine of £1000, and each was required to go to College Green and 'there to stand before the statue for half an hour, with the words: "I stand here for defacing the statue of our glorious deliverer, the late King William" inscribed on boards upon them.'[18]

While the students' attack on the statue may have owed more to the fact that they 'indulged themselves too freely in drinking',[19] and that the king sat astride his horse with his back turned to the College, other more politically motivated attacks occurred in succeeding decades. Many Catholics were angered by the rendering of annual homage to a King who had in effect been instrumental in their suppression. On 21 July and 4 November the statue was coloured white, decorated with orange lilies, a flaming cloak and sash, while the horse was decorated with orange streamers, and a bunch of green and white ribbons was symbolically placed beneath its raised foot. The railings were also painted orange and blue 'and every person who passed through College Green on these occasions was obliged to take off his hat to the statue'.[20]

Such annual displays were controversial and contentious events in the capital and produced much bitterness. In 1805, for example, on the eve of the 4 November procession, it was painted black. A year later the inflammatory nature of the annual processions was recognised by the Lord Lieutenant, the Duke of Bedford, when he refused to attend. The practice of firing volleys was discontinued shortly afterwards, although the annual decoration continued. During the mayoralty of Sir Abraham Bradley King (1820–1), an attempt was unsuccessfully made to halt the processions altogether.[21] In November 1822 the

Lord Mayor issued a proclamation prohibiting the decoration of the statue. Subsequently, the processions were abandoned, although the monument continued to be subject to attack on numerous occasions.[22] Later in the nineteenth century, just days after the unveiling of the O'Connell monument at the head of Sackville Street, (later renamed O'Connell Street), a crowd of several hundred people gathered around the statue of William III in College Green. The rioting which ensued, along with the damage to the monument, led to a number of arrests.[23]

The monument dedicated to King William III effectively served as one of the first symbols of Protestant Dublin's allegiance to the crown. Equally, however, it was a means by which those at odds with the established regime could give vent to their dissatisfaction, hence the many attacks it suffered. It also served as a prototype for the other statues erected in the eighteenth century and dedicated to members of the British Royal family, namely kings George I and George II.[24]

King George I, Essex Bridge

The accession of George I (1660–1727) to the throne generated much celebration in the capital as Protestant Dublin sought to demonstrate its allegiance to the newly instituted ruling family, the Hanoverians.[25] The king was styled as the second deliverer of the city given that his government had suppressed a Jacobite movement early in 1716. While the anniversary of his birth was celebrated in the city and his portrait was presented to the Tholsel, the fact remained that he too, like King William III, was a contentious figure. This was demonstrated in 1719 when on 29 June:

> the Guildhall of the Tholsel of this city was broke into by some person or persons disaffected to his most sacred majestie, king George, and his government; who in the night time . . . broke into the said hall and defaced and cut in pieces his said majesties picture.[26]

Nevertheless, a Dublin Corporation committee was formed in 1717 to institute proceedings to erect a monument 'in grateful acknowledgement of the many favours conferred on this city by his present majestie King George'.[27] The Dutch sculptor John Van Nost was commissioned to sculpt an equestrian statue of the king which was to be erected on Essex Bridge, one of the major route-ways between the north and south inner city.[28] His monument depicted the king in the guise of a Roman Emperor astride his horse, and wearing contemporary battle dress with high boots and spurs, and a sheathed sword at his side. It was eventually unveiled on a pier on the west side of Essex Bridge on 1 August 1722 (plate 2.2).[29] It is significant that the city assembly chose to erect the monument on Essex Bridge, a focal point and chief artery in the eighteenth-century city.

Plate 2.2 **King George I, Essex Bridge**
Erected in 1722 and sculpted by John Van Nost.
From Charles Brooking's *A map of the city and suburbs of Dublin* (London, 1728).

The figure of King George I was not destined to remain long on Essex Bridge, for the rebuilding of the bridge in the mid-eighteenth century necessitated its removal. On 19 January 1753 the statue and pedestal were removed to storage in the house of Lord Longford on Aungier Street. In 1782 it was relocated to the grounds of the Mansion House in Dawson Street. Ten years later the statue had fallen down and it is recorded that 'small injury has happened thereto'.[30] While suggestions were put forward in the city assembly that the statue be re-erected in Fitzwilliam Square, this never came to pass, although the rebellion of 1798 brought the statue renewed attention.[31] In a measure which can be interpreted as an act of defiance, it was brought from the rear garden to the side of the Mansion House overlooking Dawson Street and an inscription was placed on the pedestal which went:

Be it remembered that at the time when Rebellion and Disloyalty were the characteristics of the day, The Loyal Corporation of the city of Dublin re-elevated this statue of the First Monarch of the illustrious House of Hanover. Thomas Fleming, Lord Mayor, Jonas Paisley and William Henry Archer, Sheriffs Anno Domino 1798.[32]

It remained in this demoted position away from the public gaze throughout the nineteenth century. As Thackeray describes in his *Irish sketchbook*, 'peering over a paling, is a statue of our blessed sovereign George II'.[33] (plate 2.3). In this location the statue was relatively secure and was not subject to the same abuse that was levelled upon the figure of William III.

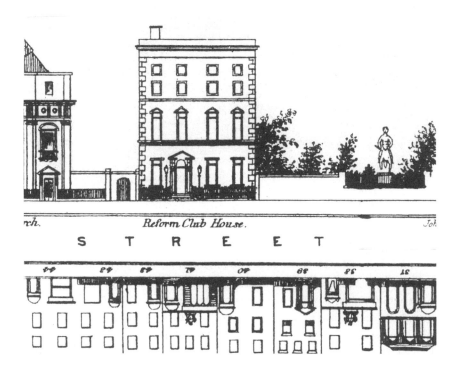

Plate 2.3 **King George I, Mansion House garden, Dawson Street**
From H. Shaw, *The Dublin pictorial guide and directory of 1850*
(Dublin, 1850, reprinted Belfast, 1988)

King George II, St Stephen's Green

But lo! A Statue from afar salutes your Eyes,
To which th' Inclosure all Access denies.
. . . 'tis Royal GEORGE on whom you stare,
Tho' oft mistaken for some good Lord Mayor:
. . . Let GEORGE's royal Form be fairly shewn,
And like his Virtues, be reveal'd and known.[34]

King William III and King George I, were joined in the city centre in 1758 by a statue dedicated to King George II (1683–1760) and erected in the then private St Stephen's Green. The first statue to be erected in the Green, it was also sculpted by John Van Nost the Younger, and depicted the king in Roman habit. In 1752 a motion was put before the city assembly, that in gratitude 'for the many and great benefits they daily enjoy under His Majesty's most gracious administration and protection . . . a statue . . . be erected . . . and placed in such conspicuous part of this city'.[35] The committee that had been established opted for a site in St Stephen's Green and made contact with the sculptor, John Van Nost, who was awarded the commission to sculpt an equestrian statue of the king.[36] The statue was placed on a tall pedestal which ensured its visibility from as far away as Nassau Street to the north and from Aungier Street to the west, for scenic as well as for defensive purposes. This was completed in 1756 and the statue was unveiled on 2 January 1758 , accompanied by a procession through the streets led by the Lord Mayor, the recorder, the aldermen, sheriffs and commons and attended by the city musicians, (plate 2.4).

The monument later became the subject of abuse as Wright's *Dublin* of 1818 notes:

> For a number of years it appeared to be destined to fall like that of Sejanus by the hands of ruffians; from its remote situation, mid-night depredators were induced to make a trial of their skill in sawing off a leg or an arm for the value of the material. One leg of the horse was cut off and a saw had nearly penetrated his neck when the watchmen were alarmed by the noise and routed the depredators.[37]

A protracted dispute arose in 1818 when the Wellington monument was about to be erected. The committee for the Wellington monument requested that the statue of George II should be removed to another site, such as Fitzwilliam Square. It was proposed, however, 'that a king ought not be removed to accommodate a subject'.[38]

Plate 2.4 **King George II, St Stephen's Green**
From the Lawrence Collection, National Library of Ireland

From monarchy to military in Dublin's public statues

The early nineteenth century was dominated by the Napoleonic Wars during which British naval supremacy and military strength were challenged by the military might of Napoleon. As might be expected, commemorative monuments to two of the heroes of this period, Nelson and Wellington, were to be erected in Dublin. These monuments served to complement the figures dedicated to the British monarchy already in place and their erection can be read as further evidence of the powerful nature of the monumental medium as a means of demonstrating the political hegemony of the Anglo-Irish ascendancy. One such monument was that dedicated to Lord Nelson erected in the centre of the city's main thoroughfare, Sackville Street.

Lord Nelson, Sackville Street

Our thought next by a towering pillar claimed,
Ascend that noble column, heavenward aimed,
Where gallant Nelson stands, with lordly grace,
To show our gratitude, and fire the youthful race; . . .
Justly he stands on that exalted pile,
Who peace preserved within our sea-girt isle;
Who barred the highway of the seas, and let us know
Only by hearing that we had a foe . . .
May thy proud statue, Nelson, and thy flag unfurled
Breathe to the winds that blow around the world,
Speak to the waves that wash our island shore,
Print on the sunbeam, ne'er impressed before,
The words that rendered Britain's sons in battle mighty,
'England expects every man to do his duty.'[39]

A monument to a hero of the Empire of which Dublin was the second city, Nelson's Pillar was erected in 1809 in the aftermath of the admiral's victory at the Battle of Trafalgar on 21 October 1805.[40] This news had been greeted with much rejoicing when it eventually reached Dublin on 8 November. Although Nelson (1758–1805) lost his life in the battle, his victory at Trafalgar, which saw the crushing of the French and Spanish fleets, ensured that Napoleon's blockade came to an end and that trade between the British Isles and the Continent could resume. With the news of Nelson's victory arose the issue of commemorating him in the city. A committee was appointed on 28 November 1805 and its members immediately set about gathering funds to pay for the monument and choosing an appropriate site. Some favoured a maritime location either at the Liffey side or at Howth Head, where it could stand by the sea as a landmark for sailors. Others argued in favour of a site in the centre of the city on Sackville Street, where it would be high enough to ensure that the figure of Nelson could look out onto the sea from his lofty perch.[41] The Sackville Street location eventually won out and the foundation stone was laid on 15 February 1808, an appropriate date which marked the anniversary of the Battle of Cape Vincent in 1797, another of Nelson's battle victories. The *Freeman's Journal* offers an account of the proceedings, which began by congratulating:

> Our country, but more particularly the metropolis on the arrival of a period, which, while it commemorates the achievements of a great naval commander,

fully evinces that the Irish people entertain as lively a sense as their fellow subjects, of the gratitude they owe to the memory of Lord Nelson.[42]

On the day that the foundation stone was laid, the streets of the city were lined with troops from Dublin Castle to Sackville Street and the grand procession started at the Royal Exchange:

> The streets were lined with military all the way up to the Rotunda, and at half past twelve o'clock, horse yeomanry, foot soldiers, sailors with flags, the Marine boys, the Hibernian School boys, the Sea Fencibles, and a host of officers of the navy and Army in uniform, formed into line, and together with the subscribers to the memorial, and a long string of private carriages, wended their way to Sackville Street. The Lord Lieutenant and the Duchess of Richmond drove the State coach, drawn by six 'of the most beautiful horses,' and brought up the rear of the procession, the members of the committee being distinguished in the centre of it by having white wands in their hands.[43]

The ceremony was led by the Lord Lieutenant, the Duke of Richmond, who was dressed in the uniform of a General and accompanied by his wife. He laid the foundation stone, and inserted a brass plate in the stone with the inscription:

> By the Blessing of Almighty GOD, To commemorate the Transcendent Heroic Achievements of the Right Honourable HORATIO LORD VISCOUNT NELSON, Duke of Bronti, in Sicily, Vice-Admiral of the White Squadron of His Majesty's Fleet, Who fell Gloriously in the Battle off CAPE TRAFALGAR, on the 21st day of October, 1805; when he obtained for his County A VICTORY over the COMBINED FLEET OF FRANCE AND SPAIN, unparalleled in Naval History; This first STONE of a Triumphal PILLAR was laid BY HIS GRACE, CHARLES DUKE OF RICHMOND and LENNOX, Lord Lieutenant General and General Governor of Ireland, on the 15th Day of February, in the year of our Lord, 1808, and in the 48th Year of the Reign of our most GRACIOUS SOVEREIGN, GEORGE THE THIRD, in presence of the Committee, appointed by the Subscribers for erecting this monument.[44]

Once the stone was laid in position three volleys were fired by the yeomanry, followed by a discharge of artillery after which all present gave three cheers while the bands played 'Rule Britannia'.[45]

With the foundation stone in place, the business of designing the monument commenced. The commission was awarded to William Wilkins, although it is thought that Francis Johnston, the architect of the General Post Office, also had a role in designing the pillar.[46] It took the form of a Doric column 134 feet in height (40.8 metres) upon which was placed a 13 ft (4 metres) statue of Nelson by Thomas Kirk, sculpted in Portland stone (plate 2.5). The names and dates of Nelson's great victories were inscribed on the four sides of the pillar, while a stone sarcophagus was placed on the southern side of the pedestal, over the Trafalgar panel. The entrance was positioned to the west side of the monument, approached by a flight of steps which brought the visitor beneath the street level under the pillar to where the spiral stairs of 168 steps commenced. The monument was eventually unveiled a year later on 21 October 1809, the anniversary of the Battle of Trafalgar.

Plate 2.5 **Nelson's Pillar, Sackville Street**
Erected in 1809 the figure of Nelson was by Thomas Kirk and the pillar by William Wilkins.
Painting by Michael Angelo Hayes, Sackville Street, Dublin, 1854, National Gallery of Ireland

Once erected on the city's central thoroughfare, the Nelson monument evoked much comment. A number of contemporary commentators rejected

the monument on political grounds, as is shown in a paragraph published in
The Irish Magazine, September 1809:

> English domination and trade may be extended, and English glory perpetuated,
> but an Irish mind has no substantial reasons for thinking from the history of
> our connexion that our prosperity or our independence will be more attended
> to, by our masters than if we were actually impeding the victories, which our
> valour have personally effected . . . We have changed our gentry for soldiers,
> and our independence has been wrested from us, not by the arms of France, but
> by the gold of England. The statue of Nelson records the glory of a mistress and
> the transformation of our senate into a discount office.[47]

Others pointed to the aesthetic shortcomings of the pillar, arguing that:

> Its vast unsightly pedestal is nothing better than a quarry of cut stone, and the
> clumsy shaft is divested of either base, or what could properly be called a capital . . .
> it not only obtrudes its blemishes on every passenger, but actually spoils and blocks
> up our finest street, and literally darkens the two other streets opposite to it.[48]

In the second half of the nineteenth century calls mounted for the removal
of the monument on the grounds that it constituted a traffic obstruction,
which is reflected in 'A Bright idea' published in the *Irish Builder* in 1876:

> Not in the centre of our city, Where the lines of traffic meet –
> In the very path of commerce, Blocking up a noble street,–
> As a figure in a picture Disproportionately tall,
> Seems to make its right surroundings Quite ridiculously small.
>
> Place it where the roaring billows Dash upon the rocky shore,
> Near where the ill-fated 'Vanguard' Shall be lifted 'never more!'
> And perchance the British navy, Gazing on their mighty chief,
> May, when overcome with – sorrow, In the ocean find relief!
>
> Then once more I do entreat you – Leave the mighty passage free!
> Take away the huge obstruction! Place the lighthouse near the sea![49]

In 1876 Dublin Corporation considered the issue of removal and it was
suggested that the pillar be re-erected in one of the city's squares,[50] but the fact

remained that the authority had no legal entitlement to remove the pillar as this power rested with the trustees. Shortly afterwards, the issue re-emerged for consideration in the British House of Commons in 1891, when the 'Nelson's Pillar (Dublin) Bill' was introduced by Thomas Sexton MP, a member of Dublin Corporation and Lord Mayor in 1888–9. The subject was hotly debated and although those in favour of the motion argued that the monument constituted a traffic obstruction, the political dimension of the issue was never far from the surface as became clear in the contentious debate that followed. The Attorney General for Ireland and the Unionist members for Ulster declared their opposition to the removal of the pillar. Macartney, a Unionist MP for South Antrim said:

> It is a monument to Lord Nelson which was erected by subscription to commemorate the valour of one of our distinguished sailors. It has been erected in one of the best thoroughfares of the City. Every Irishman ought to take an interest in its preservation and is entitled to intervene in any action that is calculated to injure it . . . I think that nothing more foolish or more futile has ever been proposed in any Private Bill brought before this House.[51]

He went on to declare that he had never felt any great personal respect for Dublin Corporation in a statement which suggested the proposal to remove the pillar may have been motivated by political bias and a reflection of the overwhelmingly nationalist composition of the municipal authority.

In response, Thomas Sexton, the MP for West Belfast, observed that:

> It is somewhat singular that a Bill which is promoted by the burgesses of the City of Dublin, in order to remove an obstruction, should find in its most active opponents two Members who represent the North of Ireland – South Antrim and South Belfast. Those who desire to remove the obstruction are those who see this pillar every day, while those who oppose the removal are gentlemen who are seldom in Dublin at all.[52]

Sexton was supported in the House by both T. D. Sullivan, the Dublin MP for College Green and T. M. Healy the MP for North Longford. The former suggested that:

> The House cannot have failed to perceive the curious circumstance that the opposition to the proposal to improve the City of Dublin comes from a little

knot of Northern Representatives. It is not a new thing to me to notice that a certain number of persons in the North of Ireland whose spokesmen are in this house, lose no opportunity of supporting everything that would tend to disfigure the City of Dublin, and of opposing everything that would beautify and improve it. No doubt it is a miserable and narrow-minded feeling, but I have seen it manifested over and over again.[53]

Healy meanwhile said that:

Although I am in favour of the Bill I sincerely hope that the House will reject it, for if an argument in favour of Home Rule is wanted it would be found in the refusal of the House to allow this matter to be inquired into by one of its own Committees.[54]

His views were supported by Webb, the MP for West Waterford who argued:

It would be better if the Bill were thrown out, because it would at once show the people of Ireland that even in such a purely domestic question as this the wishes of the citizens of Dublin are not to be consulted . . . More time has been devoted to the consideration of such a paltry question than is sometimes devoted to questions upon which the happiness and fortunes of our fellow-subjects depend.[55]

Although the Bill was eventually passed by a majority of 14, the trustees once again declared their unwillingness to see the monument disturbed and Nelson's Pillar remained in place. Although the issue then dropped from the political agenda until after the birth of the Irish Free State in 1922, the pre-Independence debate surrounding Nelson's Pillar reveals the potent symbolism of one particular public monument. While it was often argued that the aesthetic shortcomings of Nelson's Pillar necessitated its removal, if only to another location in the city, the political argument that it was inappropriate to commemorate a hero of the British military establishment in the heart of the city was often articulated. The debates which took place in the British House of Commons and the chamber of Dublin Corporation itself point to the nationalist composition of the municipal authority, which had escalated since the 1840s, and to the role of that body in attempting somehow to nationalise Dublin's monumental landscape. This was only one element of a broader process, however, and in the decades which followed the authority sanctioned sites on

Sackville Street for a range of monuments to prominent figures of Irish nationalist politics. In the meantime, the figure of Nelson was joined by an obelisk dedicated to another member of the British military establishment, the Duke of Wellington.

The Duke of Wellington, Phoenix Park

The statue of the iron Duke claims now
Our marked attention, at his name we bow,
For battles and proud victories are contained
In one word, Wellington, wherever named;
To distant generations men shall make,
His name their bulwark, when the sword they take;
His iron firmness they shall imitate,
In raging battle or affairs of state.
Ne'er had a grateful country greater cause
A monument to raise, midst loud applause;
And willingly in life and death they gave
Grateful acknowledgement to one so brave.[56]

Like Nelson, Arthur Wellesley, the Duke of Wellington (1769–1852) had played a key role in a number of early nineteenth-century military campaigns, most famously at Waterloo in 1815 where he secured victory over Napoleon and ended French domination of Europe. Unlike Nelson, however, and indeed those who had been commemorated on Dublin's streets before him, Wellington was Irish born and had been elected to the Irish Parliament in 1790.[57] It is perhaps not surprising then that his native city sought to commemorate his victories in the capital.[58]

In July 1813 a meeting was held 'of noblemen and gentlemen to discuss the idea' of erecting a monument, and a resolution was passed that 'a suitable testimonial would be provided in the metropolis of Ireland'.[59] A fund was initiated and a committee established to carry out the work of collecting further funds, while a second committee was set up to decide upon a design and site for the proposed monument. As McGregor observed in 1818, 'Some time back a voluntary subscription was entered into by many of the inhabitants of Dublin, and other parts of Ireland, to erect a monument which should commemorate the extraordinary achievements of their heroic countryman.'[60] Designs were then invited for a memorial that 'would adequately convey an idea of the character of the subject'.[61] The models were put on display in the exhibition rooms of the Dublin Society in order to give the general public an opportunity to view them. In 1815 Sir Robert Smirke's proposal was selected. His design, as Warburton *et al.* described in 1818, was for a monument:

> On the summit platform of a flight of steps, of an ascent so steep, and a construction so uncouth, that they seem to prohibit instead of to invite the spectator to ascend them, a pedestal is erected of the simplest square form, in the die of which, on the four sides, are as many panels, having figures in basserelievo, emblematic of the principal victories won by the Duke. Before the centre of what is intended for the principal front is a narrow pedestal insulated, and resting partly on the steps and partly on the platform. This pedestal supports an equestrian statue of the hero. From the platform a massive obelisk rises, truncated and of thick and heavy proportions.[62]

The question of where the monument was to be located was the next for consideration. A number of locations were suggested, among them the Phoenix Park, St Stephen's Green, a site off Dame Street facing St George's Street, the Rotunda Gardens, Mountjoy Square, College Park and Merrion Square. A site

known as the salute battery in the Phoenix Park was eventually chosen and the foundation stone was laid by the Lord Lieutenant, the Duke of Richmond, on 17 June 1817.

While the base, pedestal and the column of Wicklow granite had been completed in 1820, owing to lack of funds the monument was not completed until 18 June 1861 when four panels were inserted around the base of the monument commemorating the victories of the Duke (plate 2.6). On 18 June 1861 the 44th anniversary of the laying of the foundation stone, the plaques were laid open to view. At the same time the pedestals, which were to accommodate an equestrian statue of the king along with a number of lions, were removed and replaced instead with sloping steps.

Plate 2.6 **Wellington monument, Phoenix Park**
Completed in 1862, designed by Robert Smirke

Conclusion

By the middle of the nineteenth century the sculptural fabric of Dublin was dominated by figures which in their symbolic sub-text created a political landscape that expressed in stone Ireland's status as a constituent component of the British empire. Monuments of British monarchs and leading members of the military, like Nelson, were a feature of many other towns and cities throughout the United Kingdom and the Commonwealth countries where they confronted inhabitants on a daily basis and represented empire in the everyday domain of the cultural landscape. However, while such monumental initiatives may have inspired loyalty and served to cultivate a sense of belonging to empire, drawing citizens into closer communion with the imperial project, in Ireland they also served as tangible symbols around which demonstrations of resistance could be articulated. Unlike other truly imperial cities where such monuments and the displays of choreographed theatre that went with their unveiling proceeded smoothly and often without debate, Dublin was caught in something of a schizophrenic position which the opposition to these statues brought into sharp focus. While for some the imperial landscape that was carved out confirmed the city's status as imperial partner and second city of the Empire, for others they underlined the fact that Ireland was a colonial subordinate. These monuments contributed to the veneer of an imperial city, but they were less successful in cultivating among citizens a more long-term sense of loyalty and imperial identity. Consequently, over succeeding decades they were to encounter opposition from various strands of Irish nationalist opinion whose representatives, struggling to establish their own sense of identity, were galvanised into action. Moreover, such groups recognised the powerful role of public statuary as a medium of representation and as the nineteenth century progressed statues were erected to leading figures of the nationalist movement in Ireland creating a landscape of contested space in Dublin which forms the focus of the next chapter. Gradually the 'one-sided' symbolic landscape, from which figures drawn from an Irish sphere of influence were notably absent, was rewritten.

CHAPTER THREE

Contested identities and the monumental landscape

The emergence of a contested landscape

Dublin is connected with Irish patriotism only by the scaffold and the gallows. Statue and column do indeed rise there, but not to honour the sons of the soil. The public idols are foreign potentates and foreign heroes . . . the Irish people are doomed to see in every place the monument of their subjugation; before the senate house, the statue of their conqueror – within the walls tapestries with the defeats of their fathers. No public statue of an illustrious Irishman has ever graced the Irish Capital. No monument exists to which the gaze of the young Irish children can be directed, while their fathers tell them, 'This was to the glory of your countrymen.' Even the lustre Dublin borrowed from her great Norman colonists has passed away.[1]

The famine which occurred in Ireland during the 1840s, together with the social and economic deprivation that accompanied it, swept aside discussions of monument building in the capital. Over the course of the decades which followed, however, the one-sided nature of Dublin's monumental landscape began to attract increased attention. As a columnist in *The Nation* observed:

We now have statues to William the Dutchman, to the four Georges – all either German by birth or German by feeling – to Nelson, a great admiral but an Englishman, while not a single statue of any of the many celebrated Irishmen whom their country should honour adorns a street or square of our beautiful metropolis.[2]

Concerted attempts to commemorate figures drawn from the sphere of Irish nationalist politics began to gather pace towards the end of the nineteenth century in tandem with the burgeoning growth in the number of nationalist city councillors sitting in the chamber of Dublin Corporation. Initially, however, monuments were unveiled which were dedicated to 'safe' figures drawn from the realm of Anglo-Irish literature. While these monuments did not create much controversy, they marked an important point of departure in the evolution of the city's iconography. A variety of committees were established which set about honouring figures such as Tom Moore, the poet and author of *Moore's melodies*, a statue of whom was unveiled on 14 October 1857 in College Street. His monument was further augmented with the erection of statues dedicated to the writer Oliver Goldsmith and the orator and statesman, Edmund Burke, on the lawn outside Trinity College and facing College Green in 1864 and 1868 respectively.[3] While these developments began the process of tackling the sculptural deficiency on Dublin's streets, it was only a tentative beginning. Their commemoration was strongly supported by the Castle administration as is shown in the subscription lists and in records of those who attended their unveiling ceremonies.

The later decades of the century, however, were characterised by the emergence of statues dedicated to more overtly nationalist figures. A significant shift began to take place in the nature of the individuals to whom statues were dedicated in the capital, as well as in the locations that they were afforded vis-à-vis figures from the domain of the British administration. Increasingly, individuals who had played leading roles in the various and often contentious strands of Irish nationalist and republican politics were commemorated. Hence, the erection of monuments dedicated to Daniel O'Connell and C. S. Parnell, both of whom had been to the fore of constitutional politics during different periods of the nineteenth century. Equally, men who had led sections of the population in violent revolt and sought the creation of an independent Irish republic were honoured, among them William Smith O'Brien (table 3.1).

While the sculptural fabric of the city continued to be augmented by the commemoration of figures associated with the Castle administration, as well as the British Royal family, it is significant that these statues were afforded much less prominent locations than their nationalist counterparts. The contested nature of Dublin's urban landscape by the turn of the century stemmed, however, from both a broad-based 'British–Irish' opposition, as well as from the various internal discourses of Irish nationalism. This was demonstrated most especially at the close of the century in the debate which surrounded the

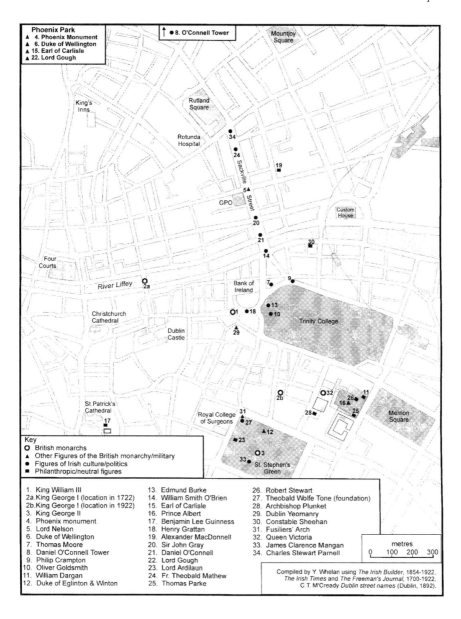

Phoenix Park
▲ 4. Phoenix Monument
▲ 6. Duke of Wellington
▲ 15. Earl of Carlisle
▲ 22. Lord Gough

● 8. O'Connell Tower

Mountjoy Square

King's Inns

Rutland Square

Rotunda Hospital

34
24
Sackville Street
19
5 ▲
GPO
20
21
30
14

Custom House

Four Courts

River Liffey 2a

Bank of Ireland 7 ● 9

Christchurch Cathedral

1 ● 18
● 13
● 10

Dublin Castle 29

Trinity College

St. Patrick's Cathedral

17

Royal College of Surgeons

2b
31
27
▲ 12
23
33 3
St. Stephen's Green

32
26 11
16 ▲
28 25
Merrion Square

Key
○ British monarchs
▲ Other Figures of the British monarchy/military
● Figures of Irish culture/politics
■ Philanthropic/neutral figures

1. King William III
2a. King George I (location in 1722)
2b. King George I (location in 1922)
3. King George II
4. Phoenix monument
5. Lord Nelson
6. Duke of Wellington
7. Thomas Moore
8. Daniel O'Connell Tower
9. Philip Crampton
10. Oliver Goldsmith
11. William Dargan
12. Duke of Eglinton & Winton

13. Edmund Burke
14. William Smith O'Brien
15. Earl of Carlisle
16. Prince Albert
17. Benjamin Lee Guinness
18. Henry Grattan
19. Alexander MacDonnell
20. Thomas Moore
21. Daniel O'Connell
22. Lord Gough
23. Lord Ardilaun
24. Fr. Theobald Mathew
25. Thomas Parke

26. Robert Stewart
27. Theobald Wolfe Tone (foundation)
28. Archbishop Plunket
29. Dublin Yeomanry
30. Constable Sheehan
31. Fusiliers' Arch
32. Queen Victoria
33. James Clarence Mangan
34. Charles Stewart Parnell

metres
0   100   200   300

Compiled by Y. Whelan using *The Irish Builder,* 1854-1922,
The Irish Times and *The Freeman's Journal,* 1700-1922,
C.T. M'Cready *Dublin street names* (Dublin, 1892).

Figure 3.1 **Statues erected in Dublin, 1700–1922**

Table 3.1 **Statues erected and proposed in Dublin, 1850–1922**

Individual	Date	Location	Sculptor
Tom Moore	1857	College Green	Moore
Daniel O'Connell Tower	1860	Glasnevin	Petrie
Philip Crampton	1862	Hawkins / College Street	Kirk
Oliver Goldsmith	1864	Trinity College	Foley
William Dargan	1864	National Gallery	Farrell
Earl of Eglinton	1866	St Stephen's Green	MacDowell
Edmund Burke	1868	Trinity College	Foley
William Smith O'Brien[1]	1870	D'Olier Street/Carlisle Bridge	Farrell
Earl of Carlisle	1871	Phoenix Park	Foley
Prince Albert	1872	Leinster Lawn	Foley
Benjamin Lee Guinness	1875	St Patrick's Cathedral	Foley
Henry Grattan	1876	College Green	Foley
Alexander MacDonnell[2]	1875	Marlborough Street	Farrell
John Gray	1879	Sackville Street	Farrell
Daniel O'Connell	1882	Sackville Street	Foley
Lord Gough	1880	Phoenix Park	Foley
Lord Ardilaun	1891	St Stephen's Green	Farrell
Fr Theobald Mathew	1893	Sackville Street	Redmond
Thomas Parke	1896	Natural History Museum	Wood
Robert Stewart	1898	Leinster Lawn	Farrell
Wolfe Tone	1898	St Stephen's Green NE	never completed
Archbishop Plunket	1901	Kildare Place	Thornycroft
Dublin Yeomanry	1904	St Andrew's Church	Murray
Constable Sheehan	1906	Burgh Quay	(unknown)
Fusiliers' Arch	1907	St Stephen's Green	Pentland and Drew
Queen Victoria	1908	Leinster House	Hughes
J. C. Mangan	1909	St. Stephen's Green	Sheppard and Pearse
C. S. Parnell	1911	Sackville Street	Saint-Gaudens

1   This statue was moved in 1929 for traffic reasons to a more prominent location on O'Connell Street.
2   This monument was removed in the 1950s following a range of attacks on statues in the city.
Sources: *Irish Builder*, 1850–1922, *The Irish Times* and the *Freeman's Journal*, 1900–22.

erection of a monument to Parnell in 1899, *before* the monument dedicated to Wolfe Tone, for which the foundation stone had been laid a year earlier, had been completed. Finally, statues dedicated to a number of philanthropic and apolitical figures were also erected on the streets of the city in the latter decades of the nineteenth century. These statues commemorated 'neutral' figures that were honoured not for any overt political reasons but as a testament to their benevolence or particular achievements. Thus Lord Ardilaun was commemorated in St Stephen's Green for his role in laying out the green as a public park in 1880, and Benjamin Lee Guinness at St Patrick's Cathedral for his work in facilitating the nineteenth-century restoration of that building. A monument was also erected to Sir Philip Crampton, the Surgeon-General and a naturalist, at the junction of Hawkins Street and College Street in August 1862,[4] while William Dargan, the main force behind the Dublin Exhibition of 1853 and the foundation of the National Gallery, was honoured with a statue in the grounds of the National Gallery in 1864. In exploring the symbolic geography of Dublin before independence and the role of public monuments in contributing to that iconography, one street in particular, Sackville Street (today O'Connell Street), merits close examination.

Sackville Street and the politics of public statuary

First laid out as Drogheda Street after Henry Moore, Earl of Drogheda, Sackville Street was redeveloped in the 1740s by Luke Gardiner, a member of one of the most important landowning families north of the river Liffey. One of his greatest achievements was the creation of Gardiner's Mall, a tree-planted walk 48 feet wide which occupied the centre of Sackville Street and which set the scale for what is now central Dublin.[5] The street, which was named after Lionel Cranfield Sackville, Duke of Dorset and Lord Lieutenant in Ireland from 1731–7 and again from 1751–5, soon became a fashionable location where the Lords and Gentry of Ireland who sat in one or other of the Houses of Parliament had their city mansions. The street was later extended by the Wide Streets Commissioners in the 1780s when Lower Sackville Street was created and Carlisle Bridge (later O'Connell Bridge) was constructed. These developments facilitated ease of access from the north side of the city to the House of Parliament and Trinity College on the south side. Over the following decades, the iconography of Sackville Street continued to evolve in tandem with the political and cultural context of the nineteenth-century city. The erection of the pillar dedicated to Lord Nelson in 1809 served as a striking symbol of Ireland's links with the British Empire. However, in the closing decades of the nineteenth century the street served a critical role in representing the discourses of Irish nationalism (see figure 3.2).

Commemorating Daniel O'Connell

Hence, amidst Erin's stately columns rise
The Liberator's monument, reaching the skies;
And if 'tis asked, why Liberator named —
For what great doings was the hero famed?
We humbly hope good reasons we can bring
To prove him in reality an Irish king . . .
A fitting monument to him we praise;
There, by the cloven column let him stand
With purse of large dimensions in his hand.[6]

The laying of the foundation stone of the O'Connell Monument in August 1864 marked the first stage in what was to become an imposing monument and a dominant political statement. The event was a momentous occasion in Dublin when the streets became thronged with people who set out to 'do justice to the memory of O'Connell. A national benefactor . . . to him the Irish people owe the liberties they enjoy – and to him the Irish people will pay to day a tribute of gratitude for these liberties.'[7] The decision to erect the monument in Dublin was preceded by the embellishment of O'Connell's grave in Glasnevin Cemetery on the northern outskirts of the city where a round tower, designed by George Petrie, acted as a counterpoint to the newly completed Wellington Monument. In 1847 the cemetery committee of Glasnevin Cemetery decided that a memorial should be erected to the memory of Daniel O'Connell. The antiquarian George Petrie came up with the design which was exhibited at the Dublin Exhibition of 1853. Petrie provided for a chapel, a high cross and a round tower. All were to be designed in the simple style of the early Christian period and placed on a raised platform. As it turned out, only the round tower was erected and O'Connell's remains were placed in the crypt beneath it in 1860. The tower became the focus of a design which was built out of proportion to Petrie's original model in which three elements, tower, cross and chapel were meant to complement one another.[8]

The origins of this Sackville Street project date back to 1862 when the O'Connell Monument Committee was established following a public meeting

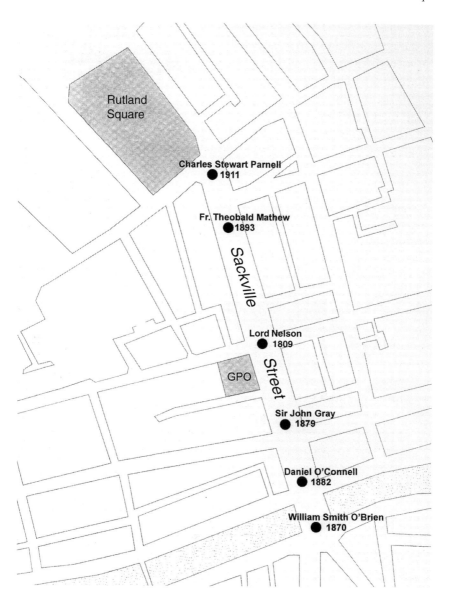

Figure 3.2 **Public monuments on Sackville Street in 1922**

in the Prince of Wales Hotel on Sackville Street.[9] A fund had been opened prior to the meeting and was heavily backed in the *Freeman's Journal* and by members of the Catholic clergy.[10] While subscriptions to their fund multiplied, the committee sought a site at the head of Sackville Street for the monument. A site at this location, within view of the former Irish Parliament House, was granted. A central committee was then established which adopted the resolution that, 'the monument would be to O'Connell in his whole character and career, from the cradle to the grave so as to embrace the whole nation'.[11]

It is noteworthy that at the same time as the O'Connell project was proceeding, a competition was launched which deflected attention away from it and towards a new bridge to replace Carlisle Bridge (now O'Connell Bridge). Among the design criteria for the bridge was the suggestion that it should provide 'statuary to do honour to the memory of illustrious men'.[12] It is of some significance that the winning design, by the Belfast architect William Henry Lynn, had a memorial not of O'Connell but of Prince Albert as its centrepiece. The decision to widen the bridge, however, was postponed, and the statue of Prince Albert was eventually erected in Dublin in a much less conspicuous location at the rear of Leinster House on Leinster Lawn in 1872.[13] With that, the proposal to erect the O'Connell monument gathered pace and the foundation stone was laid two days after the centenary of O'Connell's birthday in August 1864 (plate 3.1). This occasion marked the first important stage in the erection of the monument and brought thousands onto the streets of the capital:

> The sun shone out brilliantly and approvingly; the countless masses swayed to and fro . . . the roofs and windows of the public buildings and houses as far as the eye could reach, were filled with people. The ladies waved their handkerchiefs, the men cheered and waved their hats. Around the enclosure were gathered the magnificent banners of the trades, the bands playing their most stirring national airs. It was a tremendous scene.[14]

The crowds processed through the streets on a symbolic route-way from Merrion Square to Sackville Street, led by the committee member, Sir John Gray, and were addressed by the Lord Mayor, Peter Paul Mac Swiney. He observed that:

> The people of Ireland meet to-day to honour the man whose matchless genius won Emancipation, and whose fearless hand struck off the fetters whereby six millions of his countrymen were held in bondage in their own land . . . casting off the hopelessness of despair, the Irish people to-day rise above their

Plate 3.1 **The laying of the foundation stone for the O'Connell monument, Sackville Street**
Illustrated London News, 20 August 1864

afflictions, and by their chosen representatives, their delegated deputies, and their myriad hosts, assemble in this metropolis, and signalise their return to the active duties of national existence, by rendering homage to the dead and by pledging themselves to the principles of him who still lives and reigns in the hearts of the emancipated people.[15]

With the foundation stone in place the business of commissioning a sculptor began. A competition was launched by Dublin Corporation which took over responsibility for the monument, and a selection committee was appointed. Prizes of £100, £60 and £40 were offered and plans were issued showing the scale of the site and descriptions of adjoining buildings. The committee did, however, retain the prerogative not to award the commission if they so desired.[16] In December it passed a resolution that, 'in as much as first class artists would not send in competing designs, the principle of competition for the design could not be advantageously adhered to'.[17] Gray was consequently requested to confer with the sculptor J. H. Foley on the subject. Foley's status as Irish-born but non-resident sculptor led to many debates over the course of the following year. The *Irish Builder* noted its respect for Mr Foley but went on to comment that 'we most emphatically protest against sending £10,000 out of the country for the execution of an undertaking which, above all others, should be thoroughly national, and as the monument originated from Irish hearts, so it should be sculptured by none other than Irish hands.[18]

The competition went ahead, side by side with the negotiations with Foley and the closing date was set for 1 January 1865.[19] By that date sixty designs had been received, each of which was described in detail in the *Irish Builder* and was exhibited in the City Hall. The O'Connell monument committee met to consider the designs on 20 February, and on 21 April reached its decision that 'it is . . . with much regret that we find ourselves unable to recommend any of them for adoption by the committee'.[20] The *Irish Builder* was scathing in its criticism of the decision:

> We think that sixty British architects would not care to be told that none of their designs are 'suitable in respect of beauty, of general outline, or proportion, or of fitness for the immediate object of the monument . . .' Irish art must we think, be in a very sad condition if such severe criticism be true of the production of sixty architects.[21]

Another competition was set in motion but the committee found that once again they were unable to recommend any of the new designs. It then contacted Foley, and made a concession to popular opinion by requesting that a resident Irish sculptor would assist him in designing subsidiary figures. Foley refused to grant that request but did agree to give an Irish architect the opportunity to furnish a design, which, if he so desired, would be incorporated into his project.[22] Although three proposals were submitted none were considered suitable and Foley went ahead with his own, which was to be the subject of a protracted gestation.[23]

By 1871 the O'Connell monument remained unfinished, leaving the *Irish Builder* to observe that, 'Six years is a long time to wait from order till execution; and as the case stands no guarantee is given that the six years may not grow into twelve or more.'[24] This also gave many the opportunity once again to resurrect the issue of the native versus foreign sculptor:

> The statues of Goldsmith and Burke grace our city, and the citizens of Dublin are satisfied as to their execution, and the Smith O'Brien monument is worthy, in point of execution, of taking its place beside them or apart. These statues

have not been delayed an unusual time, and on all sides satisfaction is consequently felt. Moreover we venture to suggest that when works of art are required, and when monuments are proposed to be erected in future in Ireland that the resident Irish artist will not be overlooked.[25]

These sentiments were echoed in even stronger terms in the *Irish Sportsman*:

How is it, then, that 'no Irish need apply' to execute the statues of Irishmen, subscribed for with Irish money, and the funds administered by those who profess the extremist nationality . . . why is it that there is nothing visible of the O'Connell monument ordered several years ago . . . Irish money for Irish labour must be the cry of those who have nationality enough of a vital kind to make a steady stir in the matter.[26]

In August 1871 Foley presented a progress report to the Corporation and explained that owing to illness and pressure of work, the progress of the monument had been delayed, though he did envisage that the monument would be completed in time for the O'Connell centenary in 1875. Foley's death in 1874, however, left the committee with a monument that would not be complete in time for the O'Connell centenary on 6 August 1875. Instead, the monument was finished by Foley's assistant, Thomas Brock, who was himself formally commissioned in June 1878. The monument was designed in three sections: at the top, a statue of O'Connell; in the middle a frieze, at the centre of which was represented the 'Maid of Erin', her right hand raised pointing to O'Connell and in her left hand the Act of Catholic Emancipation; and finally four winged victories were placed around the base, each of which represented the virtues attributed to O'Connell – patriotism, courage, eloquence and fidelity. All three sections combined to record 'the gratitude of the Irish people for the blessings of civil and religious liberty obtained for their native land by the labours of the illustrious O'Connell'.[27] The figure of O'Connell was ready for unveiling at the head of Sackville Street in August 1882.

The eve of one of Ireland's greatest days has now arrived. Every element of success attended the centennial and O'Connell celebration. Numbers, strength, enthusiasm, and all the adjuncts, natural and artificial, of popular triumph wait upon tomorrow's festival . . . if the O'Connell bronze, whose heroic beauty will be revealed to the populace tomorrow, could speak, it might tell them, too, that many monster meetings of the past looked down upon them, Tara Hill and

Mullaghmast, the meetings of the Funeral, the Foundation Stone, and the centenary stand before the people for comparisons with tomorrow's.[28]

On 15 August 1882, Sackville Street acted as the theatre for a 'monster meeting' of considerable proportions. The day was marked by a procession through the capital (figure 3.3) that took the participants past a range of buildings with which O'Connell had some form of association. The monument was eventually unveiled at one o'clock, when 'a mighty roar' went up 'from ten thousand throats when the veil fell at the Lord Mayor's signal'. The committee delivered the statue over to the care of the Corporation which the Lord Mayor accepted with a few brief remarks, and 'with a quick touch withdrew the covering from the Herculean figure of O'Connell. At that instant the sun suddenly opened its beams through the drenching rain, and gloriously lighted up the Monument and the crowded platform (plate 3.2).'[29]

Politics, religion and water supply: commemorating Smith O'Brien, Gray and Fr Mathew

While the foundation stone for the O'Connell monument was laid in 1864, the monument itself was not completed until 1882 and in the intervening period a number of other notables were commemorated in prominent locations in the capital. Among them was the statue dedicated to William Smith O'Brien, a leader of the doomed rebellion of 1848 (plate 3.3). This was the first monument erected in Dublin to commemorate an individual who had stood for armed resistance to British rule. It was followed by a statue dedicated to Sir John Gray, another prominent nationalist, and owner of the *Freeman's Journal* newspaper (plate 3.4). Finally, a statue of the Apostle of Temperance, the Capuchin Friar, Fr Theobald Mathew, was unveiled in 1893 (plate 3.5).

The date chosen for the unveiling of a monument dedicated to William Smith O'Brien (1803–64), a descendant of the Protestant nobility who traced his lineage back to Brian Boru, was 26 December 1870. As he had been a revolutionary nationalist, the statue of Smith O'Brien broke the sculptural mould in the capital. The occasion was a significant one for it marked:

the first time for 70 years that a monument had been erected in a public place in Dublin to honour an Irishman whose title to that honour was that he devoted his life to the Irish national cause (Cheers). In other countries it is such men only that received the honour of a public monument, but in this city there were statues to men who had served and loved England, and did not care for Ireland.

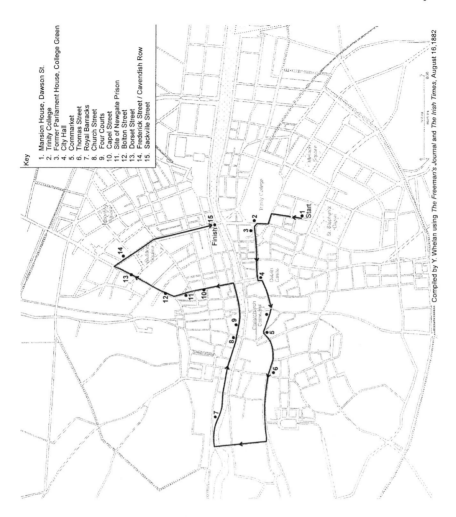

Key

1. Mansion House, Dawson St.
2. Trinity College
3. Former Parliament House, College Green
4. City Hall
5. Cornmarket
6. Thomas Street
7. Royal Barracks
8. Church Street
9. Four Courts
10. Capel Street
11. Site of Newgate Prison
12. Bolton Street
13. Dorset Street
14. Frederick Street / Cavendish Row
15. Sackville Street

Compiled by Y. Whelan using *The Freeman's Journal* and *The Irish Times*, August 16, 1882

Figure 3.3 **The processional route for the unveiling of the O'Connell monument, August 1882**

Plate 3.2 **Daniel O'Connell, Sackville Street**

Plate 3.3 **William Smith O'Brien, Sackville Street**

Plate 3.4 **Sir John Gray, Sackville Street**

Plate 3.5 **Father Theobald Mathew, Sackville Street**

As to this country, it had been held that it was treason to love her, and death to defend her. The monuments which had been erected till now have been rather monuments of this haughty mastery of the English people and our servility and helplessness. A favourable change took place recently. Ireland had ventured to erect statues to Moore, Goldsmith and Burke, whose genius was Irish, and whose sympathies also were mainly Irish. Though these men loved Ireland, and their memories were thus commemorated, none of them ever exposed themselves to the danger of imprisonment or transportation for life for Ireland. There stood the statue of a man who 22 years ago, was sentenced to be hanged, drawn, and quartered for his love of Ireland.[30] (Cheers)

O'Brien had been sentenced to death for high treason resulting from his part in the insurrection of 1848.[31] Soon after his death in 1864, a committee was formed in order to gather subscriptions and organise the erection of a monument in his honour and, 'To this undertaking men widely differing in their political and religious sentiments have subscribed, desiring to testify their respect for the noble and honourable character of our distinguished country man, whose unselfish devotion to, and sacrifices for, Ireland have never been questioned even by the sternest critics or severest censors.'[32] The committee was led by John Martin and John Blake Dillon, both of whom had been caught up in the radical politics of the mid-nineteenth century and shared O'Brien's belief in physical force. They commissioned one of the most prominent sculptors of the day, Thomas Farrell, to sculpt the figure of O'Brien.[33] Farrell sculpted a marble figure in 'an ordinary frock coat, high buttoned waistcoat and pantaloons, all of which are treated with the most commendable taste and skill. There is not the slightest approach to stiffness in the pose which is most easy and natural.'[34] Smith O'Brien was positioned with arms folded as if addressing an assembly and in the stance of one commanding attention of a crowd. Farrell 'selected what must be acknowledged a suitable site . . . the space halfway between Carlisle Bridge and D'Olier Street'.[35]

The statue was unveiled on 27 December 1870 with the inscription:

William Smith O'Brien,
 born 17th October 1803,
sentenced to death for high treason
on the 9th October 1848.
Died 16th June 1864.[36]

The authorities, sensitive to O'Brien's politics, prohibited the processions that usually went hand in hand with the unveiling of a monument, and forbade the bands from playing music while marching.[37] The order did not deter the masses who assembled in huge numbers, however, as they had done for the laying of the O'Connell foundation stone. Bands took up their positions round the site of the O'Connell monument, near the O'Brien monument and in the adjoining streets where they continued playing throughout the proceedings.

Plate 3.6 **The emerging monumental landscape on Sackville Street**
The statue of William Smith O'Brien can be seen in the foreground while that of Sir John Gray is visible between the O'Connell monument and Nelson's Pillar.
From the Lawrence Collection, NLI

The unveiling of the Smith O'Brien monument and the displays of nationalism that went with it signalled a change in the sculptural composition of the city which was to be further reinforced with the unveiling of a monument dedicated to Henry Grattan at College Green in 1876. As was pointed out at the unveiling of the Smith O'Brien statue:

Soon two more statues would stand in prominent places in the city – the sculptured effigy of Henry Grattan before what was our Parliament House as if waiting for the re-opening of its doors to a restored Irish legislature – and the monument to the liberator of his Catholic fellow-country-men, Daniel O'Connell. These statues would provide what it was that the future of Ireland was expected to be.[38]

The broadly nationalist complexion of the monuments on Sackville Street was further reinforced with the unveiling of the monument dedicated to the nationalist MP Sir John Gray in 1879. Gray had died in 1875 and little time was wasted before establishing a committee to erect a statue to the man who, as a moderate nationalist, had played a key role in the introduction of a water supply to Dublin from County Wicklow in 1868. As chairman of the Dublin Corporation waterworks committee from 1863 until his death, he sought to give the city and suburbs an efficient water supply. A site for a monument in his honour was granted by the Corporation in 1877 again on Sackville Street close to the Abbey Street offices of the newspaper the *Freeman's Journal*, of which Gray had been editor.

The monument committee approached Thomas Farrell to design the monument. He represented Gray 'in the guise of a Victorian gentleman, complete with open coat, confident stance and a serious yet kindly expression'.[39] The statue was unveiled on 24 June 1879 with the inscription:

> Erected by public subscription to Sir John Gray Knt. MD JP, Proprietor of The Freeman's Journal; MP for Kilkenny City, Chairman of the Dublin Corporation Water Works Committee 1863 to 1875 During which period pre-eminently through his exertions the Vartry water supply was introduced to city and suburbs Born July 13 1815 Died April 9 1875

The erection of these monuments in the 1870s and 1880s marked a significant turning point in the evolving symbolic geography of Dublin's monumental landscape, further reinforced with the erection of a monument dedicated to the 'Apostle of Temperance', Fr Theobald Mathew, in 1890. Sculpted by Mary Redmond, the foundation stone for this monument was laid on 13 October 1890, the centenary of Mathew's birthday. After Dublin Corporation granted permission for the Sackville Street site a motion was passed which requested that:

the Council approve of the Right Hon. the Lord Mayor publishing a proclamation to the citizens of Dublin for a general public holiday to be kept by them during Monday 13, for the due celebration of the Father Mathew memorial proceedings upon that day, and especially that all houses licensed for the sale of intoxicating liquors be kept closed. [40]

The motion was ruled out of order, however, as it did not appear on the summons for the meeting, although the Lord Mayor did point out that while the council could not compel anyone to close his house, 'no doubt the expression of opinion in favour of the proposals would have due effect'. The monument was formally unveiled three years later with a simple inscription, 'The Apostle of Temperance, Centenary Statue, 1890, 1893 unveiled'. Sackville Street was also to be the location for one of the last sculptural initiatives in the city before independence when, in 1899, the foundation stone was laid for a monument dedicated to Charles Stewart Parnell.

Remembering C. S. Parnell

The decision to erect a statue to one of the leaders of constitutional nationalism in nineteenth-century Ireland gathered momentum in the final years of the century, spurred on by contemporaneous moves to commemorate W. E. Gladstone. The Gladstone national memorial fund had been established in 1898 when it was proposed that three monuments be erected in his memory, one each in London, Edinburgh and Dublin, three pillars of the Empire. The suggestion did not meet with much support in Dublin, however, especially since no monument had been erected by then in memory of Parnell. As the *Irish Independent* put it:

> Is it to be tolerated that such a man should have one of the prominent public places in the capital of Ireland for a statue . . . a daring insult to the memory of Parnell?[41]

Dublin Corporation echoed these sentiments when it passed a motion that:

> No statue should be erected in Dublin in honour of any Englishman until at least the Irish people have raised a fitting monument to the memory of Charles Stewart Parnell.[42]

The Gladstone proposal was subsequently dropped and the statue planned for Dublin was later erected in Hawarden, Essex, in 1925.[43] In the meantime Parnell's successor, John Redmond, set about the business of erecting a memorial to his former leader with the intention of reuniting the party. The proposal was a controversial one, not least because the Wolfe Tone monument had not been completed. The decision to erect a statue to Parnell gathered momentum and the foundation stone was laid on 8 October 1899 at the north end of O'Connell Street, eight years after his death and amid scenes of tumult and disorder. As a report in the *Freeman's Journal* observed:

> Whatever view Irishmen may take regarding Parnell's attitude during the last months of his life, they have but one concerning the value of his services to the nation while he was the trusted representative of a united people. Those services were great; they might have become priceless had they been crowned, as it was possible to crown them by the conquest of a native legislature. But though lacking its great completion, Parnell's career and the purpose that so long inspired it are worthy of a national commemoration at the hands of the Irish race.[44]

Moreover efforts to disrupt the stone laying and embarrass the organisers were taken by the Irish Republican Brotherhood (IRB). When the monument was eventually unveiled in 1911 these tensions had all but evaporated.

> Dublin was yesterday the scene of one of the most remarkable episodes in its annals. The unveiling of the Parnell monument would of itself be a memorable event, certain to attract worldwide interest. All the circumstances surrounding the event have, however, tended to emphasise its interest and importance. The political outlook is eminently favourable to the Irish cause. Never in the history of the constitutional movement has the atmosphere been clearer and brighter. So much is this the case that even the inveterate enemies of the cause of Irish freedom have made up their minds that beyond playing a game of bluff during the next year or two, they can do nothing to obstruct the establishment of a sound system of national self government in Ireland . . . if we are on the threshold of Home Rule, no man is more accountable for that fact than Parnell.[45]

At the twentieth anniversary of Parnell's death on 1 October 1911, the unveiling ceremony for the monument dedicated to him drew a large, enthusiastic crowd similar to the Wolfe Tone and O'Connell celebrations of some years earlier.

> There must have been many in the huge concourse who were able to recall that memorable day in August twenty-nine years ago when the O'Connell monument was unveiled. Until yesterday that demonstration was regarded as the greatest Dublin had ever seen. Yesterday's, in point of numbers and in representative

character, seems to have been quite equal to anything witnessed in our city within living memory. No element was missing that could have lent interest and emphasised the national character of the demonstration. Every part of Ireland was represented.[46]

The ceremony was preceded by a procession along the route of which 'the streets were lined with people and in nearly every window and balcony, and even on the house tops in many places, interested and cheering spectators were to be seen.'[47] The ceremony differed radically from the contentious scenes that had characterised the laying of the foundation stone twelve years previously. The presence of over fifty MPs together with some members of the Catholic clergy clearly indicated the changed environment, while all parts of nationalist Ireland were gathered for the event, even though a rail strike on the Great Southern and Western Railway prevented a number of groups travelling from the south. The unveiling ceremony was preceded by a procession along the route of which 'the streets were lined with people and in nearly every window and balcony, and even on house tops in many places, interested and cheering spectators were to be seen'.[48] Redmond officiated at the ceremony and in his oration he emphasised the historical significance of the occasion, marking he said the beginning of a new era in which Parnell's long awaited goal for Ireland was on the verge of reality. He went on, 'as certain as any human thing can be . . . there would be an Irish parliament assembled in this metropolis within four and twenty months'.[49] He then proceeded to perform 'the proudest action of my life pulling the cord and unveiling this noble monument, the product of the greatest genius of the greatest sculptor of his time, himself the son of an Irish mother to the memory of the greatest son of Ireland since the days of Hugh O'Neill'.[50]

The monument was designed by the Dublin-born American, Augustus Saint-Gaudens and took the form of a triangular shaft of Shantalla (Galway) granite against which a statue of Parnell was placed. The figure was sculpted in the guise of a statesman, clothed in a frock coat and in the act of speaking, with his right arm extended from his body. The figure was positioned standing by a table over which was draped a large flag of Ireland. Both statue and table were designed to stand against the broad base of the shaft which was crowned with a bronze tripod (plate 3.7). The names of the four provinces and the 32 counties of Ireland were featured, along with a number of other classical motifs including ox skulls and swags to decorate the base. A harp was also incorporated into the design together with an inscription:

Plate 3.7 **C. S. Parnell, Sackville Street**   *Plate 3.8* **Parnell monument, inscription**
Erected in 1911, sculpted by Augustus Saint-Gaudens

No man has a right to fix the boundary to the march of a nation. No man has a right to say to his country – thus far shalt thou go and no further. We have never attempted to fix the ne plus ultra to the progress of Ireland's nationhood and we never shall.

The erection of these monuments on Sackville Street marked a significant turning point in the evolving symbolic geography of Dublin's monumental landscape. In terms of the individuals which these monuments sought to commemorate, the geographical positions which they were afforded and the choreography of the unveiling ceremonies, they are indicative of a broader change in the politics of power in late nineteenth century Dublin. The monuments erected on Sackville Street, while perhaps the most prominent geographically, were not alone in making tangible the contested political context that prevailed. In fact they were joined by monuments dedicated to the parliamentarian who gave his name to 'Grattan's parliament', Henry Grattan, the writer James Clarence Mangan, and by a flurry of monumental activity that accompanied the centenary of the 1798 rebellion.

Other monumental projects of the late nineteenth century

Commemorating Henry Grattan, College Green

The monument dedicated to Henry Grattan was unveiled on 7 January 1876 outside the former Parliament House on College Green in which he had spent so much of his working life. In 1869 a committee was established and it granted the commission to John Henry Foley. Foley completed the monument in 1873 after some delay.[51] The statue represented Grattan in 'one of his characteristic attitudes while speaking, his right hand and arm being stretched out, while his left hand is placed on his breast . . . intense emotion is stamped on his face'.[52] The unveiling brought tens of thousands to College Green, where a report of the committee was presented by the Secretary J. P. Vereker and a number of MPs read short addresses to the gathered assembly. The statue was then formally handed over to the Lord Mayor and to the care of Dublin Corporation.[53] The day was once again marked with much celebratory display.

> The interesting ceremony of yesterday took place at College Green, in the presence of the citizens of Dublin and people of the surrounding districts, and who might be counted by tens of thousands. But these did not represent all who wished to do honour to the illustrious Grattan, for, along the entire route by which the Trades' procession passed, the streets were crowded, and the windows were filled with spectators of all classes, who testified, in a marked degree, that their sympathies were with the ceremonial of the day . . . In the statue unveiled yesterday Irishmen possess an effigy of Henry Grattan 'in his habit as he lived', which shall be an enduring monument of the testimony borne by all classes of his countrymen to the patriotic worth, statesmanlike ability, and great public and private virtues of the foremost orator of his age . . . The idol of the Irish people while they still had a national parliament.[54]

Plate 3.9 **Henry Grattan, College Green**

The article went on to note the suitability of the site, close to where the volunteers of the 1798 rebellion had gathered, adjacent to the Parliament House with which Grattan had been so much associated and in front of Trinity College where he had been educated. Vast crowds assembled throughout the morning of the unveiling and occupied 'all coigns of vantage for the purpose of witnessing the procession and ceremonial – all the public buildings were crowded, a platform was erected close to the statue for invited guests. The day was marked by a trades procession which mustered at the Custom House at midday and marched via Gardiner Street to the College Green.'[55]

Remembering 1798 in 1898: The Wolfe Tone foundation stone, St Stephen's Green

As it stood on the threshold of the twentieth century, the sculptural elements of Dublin's urban landscape comprised a blend of sculptured forms that betrayed the fractured and contested nature of politics in the country at large. Statues were in place which represented in symbolic form Dublin's position within the British Empire, while at the same time statues were being erected which reflected the nationalism of the municipal authority and the desire for independence. Two commemorative events which took place towards the close of the century underlined this division. The first was the Diamond Jubilee celebrations in honour of Queen Victoria during the spring and summer months of 1897. The government and the loyalist population staged civic celebrations, elaborate church services, dress balls, concerts, and the principal Irish cities were illuminated and bonfires raged throughout the country.[56] A year later the centenary of the 1798 rebellion was celebrated in the city. In terms of the numbers killed, foreign involvement, geographical extent and number of participants, the 1798 rebellion was an unparalleled revolutionary event. Its commemoration was marked throughout the country with the erection of monuments dedicated to those who had been killed.[57] The centenary provided an opportunity for nationalists to commemorate what they saw as Ireland's loyalty not to Britain but to Irish heroes and to the struggle against British rule.[58]

In the city the occasion was marked with the laying of a foundation stone to a leader of that rebellion, Theobald Wolfe Tone, at the northwest corner of St Stephen's Green.

Yesterday, favoured by all the most propitious elements and attended by an almost unprecedented gathering of people of all shades of Nationalist opinion, the demonstration in honour of Wolfe Tone took place in the city. The occasion will certainly be forever a memorable one in the annals of Irish nationality. Never since the famous stone upon which the O'Connell monument was laid has there been anything within measurable distance of comparison with the vastness and greatness of the celebration.[59]

'Wolfe Tone day' as it was referred to, 15 August 1898, was the day set aside for the laying of the foundation stone of the monument dedicated to Tone. The event brought out onto the streets of the capital the largest public gathering since the unveiling of the O'Connell monument in 1882, with an estimated gathering of 100,000 people. The day began at Cave Hill in Belfast where the foundation stone for the monument originated. This was significant as local nationalists had been prevented from erecting a Tone monument themselves in the city, so, by donating the foundation stone they got some share in the '98 commemoration. It was quarried, finished and inscribed with the words: '1798. Tribute to Theobald Wolfe Tone, patriot. Belfast nationalists '98 centenary association'. The stone was then paraded through Belfast on a decorated lorry before being loaded onto a train and sent to Dublin.[60] On reaching Dublin the stone was met by a delegation of the National Centenary Committee and taken to the site of the old Newgate Prison on Harcourt Street where it was 'laid in state' alongside another foundation stone dedicated to other patriots who had died in that prison. The laying of the stone in Dublin was marked by a mammoth procession which started at Rutland Square in the north inner city and was composed of dignitaries, up to eighty bands and costumed figures on horseback. The marchers and spectators all wore badges of green, white and orange, while flags, banners and decorative arches of evergreen lined the route, and for the first time in many years a green flag was attached to the top of Nelson's Pillar. At the head of the procession was the foundation stone of the monument.

The procession took three hours to cover the three-mile journey and brought its participants through crowded streets in a symbolically rich route, each stop chosen for its associations with the United Irishmen and the rebellion of 1798. Figure 3.4 maps the route taken by the processionists and marks the locations at which they paused as they wound their way on a circuitous route from Rutland Square to St Stephen's Green.[61] At the northwest corner of the Green the dedication rites were carried out with numbers reported to have reached one hundred thousand.[62] The ceremony was performed by the old Fenian John O'Leary who drew attention to the various attempts to achieve Irish freedom and laid the foundation in place using a trowel sent from America by Tone's granddaughter. He tapped it into place six times, one for each of the four provinces, once for the United States and once for France. The ceremony drew to a conclusion with the singing of the popular ballads, 'The memory of the Dead' and 'Who fears to speak of '98' ringing out.[63]

The site chosen for the monument was significant. A number of possibilities had been considered by the organising committee including a provocative location facing the main guard gate of Dublin Castle, at the junction of Lord Edward Street and Cork Hill. They opted, however, for the St Stephen's Green location which was close to College Green, and lay in the heart of an area popularly seen as the hub of unionism. By raising a statue to Wolfe Tone in what was seen as the most English part of the capital, a potent message was delivered. Although the monument was never completed, it was intended not as a symbol in bronze of the sorrows of Erin, but rather, '[it] will be typical of all that is combative in our race. It will be the figure of a soldier of freedom, erect and proud, the embodiment of all that is courageous and bold in a nation that has borne more sorrows and suffered more injuries than any other, and – lives.'[64]

J. C. Mangan, St Stephen's Green

On 22 May 1909, the monument dedicated to the Irish poet, James Clarence Mangan, was erected in St Stephen's Green by the National Literary Society. The unveiling ceremony was attended by many with an interest in the art and literature of Ireland, and was directed by Dr George Sigerson, the president of the society who paid tribute in his address to 'a really great soul, who, against the dark background of his life, had raised a fabric of fair poetry which was the admiration of other peoples and the glory of his own'.[65] He continued:

> In the name of the National Literary Society of Ireland, I now unveil and confide to the custody of the Commissioners and to the care of the public this memorial of Clarence Mangan. Against the dark background of his life he raised a fabric of fine poetry, which shines bright as 'apples of gold amid foliage of silver'. . . In gratitude for his genius, in memory of his patriotism, in evidence that our generation is not forgetful of benefactors, and in the hope of inspiration to future times, we erect this monument. Here in the city of his birth, in the land of his love, we erect it, bearing its beautiful symbol of our ideal Erinn, whose desire and whose honour abide in the noble affection of an undivided nation.[66]

A recitation of Mangan's poem, 'The nameless one', 'was listened to with rapt attention and was loudly applauded on the close', while this was followed

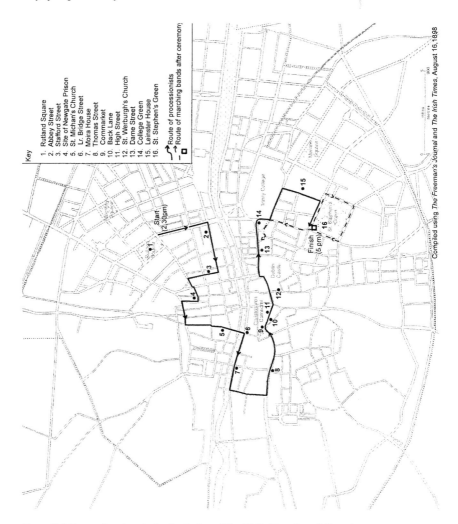

Key

1. Rutland Square
2. Abbey Street
3. Stafford Street
4. Site of Newgate Prison
5. St. Michan's Church
6. Lr. Bridge Street
7. Moira House
8. Thomas Street
9. Cornmarket
10. Back Lane
11. High Street
12. St. Werburgh's Church
13. Dame Street
14. College Green
15. Leinster House
16. St. Stephen's Green

— Route of processionists
--- Route of marching bands after ceremony

Figure 3.4 **Processional route for the laying of the Wolfe Tone foundation stone**

with speeches from Professor Gwynn,
Mr. John [Eoin] MacNeill, the vice-
president of the Gaelic League, who
addressed the assembly in Irish, and
Mr T. D. Sullivan. Gwynn noted
how it had taken some sixty years
since Mangan's death before a monument was erected in his memory and
went on to allude to the distinctively
'Irish' nature of his poetry: 'there was
a rhythm in his verses that was not
like the rhythm of any other English
poetry. Mangan must have got that
beautiful rhythm from the ancient
poetry of his fatherland, and that was
why his poetry seemed exotic to those
whose ears were attuned to different
music . . . it mattered not whether the
English public accepted Mangan's
poetry, or not, so long as his own
people accepted him and gave him a
home deep in their bosoms.'[67]

Plate 3.10 **James Clarence Mangan,
St Stephen's Green**

The monument was sculpted by the Dublin sculptor Oliver Sheppard who
had played a key role in sculpting the 'Pikemen' monuments for the centenary
of the 1798 rebellion. It took the form of a bust of Mangan, resting on a
limestone pedestal standing to a height of nine feet and also contained a niche
in the pedestal with a bust by Willie Pearse.[68] It bore the inscription:

James Clarence Mangan,
Born May 1st 1803
Died June 20th 1849
Erected by the National Literary Society of Ireland

The erection of these monuments and the often elaborate ceremonies that
went with their unveiling can be interpreted as an attempt to challenge the
ideology represented in earlier erected statues that had been unveiled in the
eighteenth century and which had become both outdated and outmoded in
the changing political and cultural context of the late nineteenth century. The

changing nature of the monumental landscape also reflects the increased power and nationalist complexion of Dublin Corporation. After all, it was the corporation that invariably granted permission for statues to be located on particular sites. Its willingness to sanction the erection of nationalistic monuments was matched by its reluctance to afford similar locations for statues that had a more loyalist bent, and which invariably ended up in more geographically isolated parts of the city.

Representing links with Empire

The sculptural map of Dublin at the turn of the twentieth century reveals that monuments dedicated to figures associated with the British monarchy, military and the Castle administration were also erected in the latter half of the nineteenth century. In an interesting shift of geography, however, they were to a large extent banished to relatively peripheral, green space locations where they were unveiled with a distinct lack of ceremony and public attention. Monuments were erected and dedicated to the Earl of Eglinton and Winton in the then private St Stephen's Green, to the Earl of Carlisle and Lord Gough in the Phoenix Park and to the Prince Consort, Albert, in the rear garden of the Royal Dublin Society at Leinster Lawn.

The Earl of Eglinton and Winton, St Stephen's Green
In 1866 a statue of Archibald William Montgomery, the 13th Earl of Eglinton and Winton, Lord Lieutenant, 1852–3 and 1858–9, sculpted by Patrick MacDowell, was erected in the then private St Stephen's Green.[69] The committee for the erection of the statue had been established in January 1862 at a meeting convened by the Lord Mayor and was attended by 'numerous parties of all sects and denominations'.[70] A subscription list was formed and 'the warmest expression of admiration for the deceased nobleman's amiable qualities and friendly feelings towards Ireland were elicited'.[71] Donations were subsequently made and by January 1862 were adjudged 'most encouraging'.[72] In March the subscriptions had risen to £6,000 and included a £200 donation from 'the Emperor of the French'.[73] The commission was granted to MacDowell and in early 1866 a site was decided upon, 'within the enclosure on the north side of St Stephen's Green, facing Dawson Street with the consent of he commissioners'.[74] The statue was erected opposite the St Stephen's Green Club of which Eglinton was a member.[75] It was unveiled in September 1866 and carried only a small notice in the *Irish Builder.*

The statue of the late earl of Eglinton, lately erected within the rails on the north side of St Stephen's Green, has been unveiled. It bears the following inscription: – Archibald William, earl of Eglinton and Winton, KT., Lord Lieutenant of Ireland, 1852, 1858–1859.[76]

The location of the statue, in what was still a private green area stood in marked contrast to the more prominent locations afforded to contemporary statues such as those dedicated to Moore, Goldsmith and Burke, and indeed the foundation stone to O'Connell. This trend was echoed in the erection of a monument dedicated to the Earl of Carlisle.

The Earl of Carlisle, Phoenix Park

The statue erected in memory of George William Frederick Howard, the 7th Earl of Carlisle, who had served as Chief Secretary from 1836 to 1841 and as Lord Lieutenant for two periods, 1855–8 and 1859–64, was erected in 1870 in the People's Gardens of the Phoenix Park. Although connected with the British administration in Ireland, it is worth noting that the Earl had given his support to the Catholic Emancipation movement and had been a part of a particularly constructive government. His statue was sculpted by John Henry Foley who represented Carlisle in the robes of the Grand Master of the Order of St Patrick (see plate 3.11). This 'meant presenting him as a member of an exclusive Protestant society rather than as a benefactor of Ireland which would in turn have had wider appeal'.[77]

The unveiling ceremony on 2 May 1870, officiated by the Earl Spencer, was a low-key affair. This may have been prompted by the fear of political demonstrations particularly when it is borne in mind that just seven months later some 20,000 people would throng the streets of Dublin for the unveiling of the O'Brien monument. *The Irish Times* noted of the monument that, 'the attitude is dignified, and as a work of art the statue is not unworthy of the known fame of the artist'. The figure was placed upon a granite pedestal and an inscription was inserted on a marble slab in the centre of the pedestal, which read:

George Wm. Frederick, seventh Earl of Carlisle, K.G..,
Chief Secretary for Ireland, 1835 to 1841;
Lord Lieutenant of Ireland, 1855–1858, and 1859 to 1864.
Born 1802. Died 1864.[78]

Plate 3.11 **The Earl of Carlisle, Phoenix Park**
Erected 1870, sculpted by John Henry Foley
From the Lawrence Collection, NLI

Albert, Prince Consort, Leinster Lawn

Will Dublin be behind-hand in erecting a memorial tribute to the deceased Prince Consort, whose virtues and attainments the Irish people so universally admired, is a question we must specially ask, when viewing the forward movements elsewhere in that regard? London, of course – Manchester, Coventry, and other important English towns speedily following the example assumed the initiative. Is it reasonable, therefore, that we should be wanting in substantially realising our professions and resolutions implying condolence with Her Majesty? Would not the area at rear of the Royal Dublin Society House, between its Museum and the National Gallery, which, under any circumstances, will most probably be decorated with statuary, form a fitting site for the purpose?[79]

The project to erect a statue of Prince Albert in Dublin began shortly after his death in 1861. Just as committees were established throughout the United Kingdom, Ireland was no different. The Dublin committee was chaired by the Lord Mayor and the commission for the statue was given directly to John Henry Foley in 1862 after which controversy arose over whether the statue should be erected in St Stephen's Green, Leinster Lawn or College Green.[80] The *Irish Builder* took a dim view of the erection of the statue in Leinster Lawn. A columnist observed that, 'if the society [Dublin Society] desire to beautify its lawn . . . it ought to look to its own exchequer for the wherewith, but not to the united contributions of the general public, against which it will surely close its gates, allowing only a peep through a gaunt, tasteless railing into the *sanctum sanctorum*.'[81] While the Leinster Lawn site had strong associational grounds with Prince Albert, given his role in promoting the Dublin Exhibition of 1853, it was a relatively secluded location. The committee opted, therefore, for a site on College Green. However, objections from nationalists who advocated the erection of a monument dedicated to Henry Grattan won out, and the Albert Committee was granted the site on Leinster Lawn.[82]

Foley's monument was to have been unveiled in June 1872. By that time, however, it remained in model form only. The committee issued a statement that owing to circumstances beyond their control they were unable to present the memorial in a complete state.[83] The unveiling of a temporary monument went ahead nonetheless, 'in order not to disappoint, if possible, public expectation, and not to deprive the visit to Dublin of his Royal Highness the Duke of Edinburgh of one of its principal objects'.[84] The committee set up a temporary pedestal, complete with a copy of the figure of the Prince Consort, so that 'His Royal Highness the Duke of Edinburgh and the public will thus have an opportunity of inspecting the memorial as nearly as possible in the exact form in which it will appear when completed' (plate 3.12).[85] When eventually finished, the monument displayed the Prince in standing position in the robes of the Knight of the Garter, the highest honour of the realm awarded by the Queen. The pedestal was flanked by four young male figures symbolising the areas of life which the Prince had promoted: art, science, manufacturing and agriculture.[86]

Lord Gough, Phoenix Park

In February 1880 a statue dedicated to Field Marshal Viscount Gough was erected in the Phoenix Park, close to the residence of the Viceroy.[87] The erection of a statue dedicated to the Field Marshal who served in the British army had first been mooted in 1869, with the commission eventually granted to Foley. He designed an equestrian statue of Gough 'straight-backed and composed, a commanding figure, an equal match for the spirited horse on which he

Plate 3.12 **The Duke of Edinburgh inspecting the Prince Consort memorial**
Erected 1872, sculpted by John Henry Foley
From *Illustrated London News*, 15 June 1872

sat astride'.[88] The unveiling of the statue was hampered, however, by the failure of the committee to find a suitable site. Initially the committee requested a site in Foster Place, close to the former Parliament House. Dublin Corporation granted this location in 1872. However, the committee opted instead for a site on Carlisle Bridge.[89] This was vetoed by members of the O'Connell committee who argued that it would spoil the view of the proposed O'Connell statue. At one point it was thought that the statue would actually end up in London as 'there has been a most unseemly squabble on the head of the site'.[90] The secretaries of the memorial committee wrote to the Corporation informing it that 'in deference to the wishes of Lord Gough and his family, and consequent on the action taken by the Corporation in withdrawing the site adjacent to Westmoreland Street and Carlisle Bridge, granted by them in 1878 for erecting thereon the equestrian statue of the late Viscount Gough, the memorial committee regret very much they are prevented placing so fine a work of art within the city of Dublin.'[91] The action of the Corporation was defended by one of its members, Alderman Harris, who complained about 'the autocratic, high-handed, and impracticable manner in which the committee have acted, and it alone is responsible for depriving the city of Dublin of this *chef d'oeuvre* of the distinguished Irish sculptor to the memory of a great and illustrious general, the late Field Marshal Viscount Gough'.[92]

By November the committee had reached the decision to erect the monument in the Phoenix Park, despite a motion passed by the Corporation which granted the Sackville Street site. A resolution was passed by the committee 'That the site in the Phoenix Park asked for from the government for the Gough memorial statue having been granted, and appearing to be most desirable . . . the committee cannot accept any other site, and consider the one now offered by the Corporation of Dublin unsuitable from an artistic point of view.'[93] The statue was finally unveiled on the periphery of the city and in a location that was closely associated with the working of the British administration – the Viceroy's residence – on 21 February 1880 (plate 3.13). The inscription on the monument was indicative of Gough's military exploits:

In honour of Field Marshal Hugh Viscount Gough, KP, GCB, GCSI, an illustrious Irishman, whose achievements in the Peninsula war in China and in India, have added lustre to the military glory of this country which he faithfully served for 75 years. This statue (cast from cannon taken by the troops under his command and granted by parliament for the purpose) is erected by his friends and comrades.[94]

Plate 3.13 **Lord Gough, Phoenix Park**
Erected 1880, sculpted by John Henry Foley
From the Lawrence Collection, NLI

Plate 3.14 **Fusiliers' Arch, St Stephen's Green**

In the years immediately preceding independence a number of other monuments were unveiled in the capital, among them two monuments dedicated to those who had fought on the British side in the Boer War at the turn of the century. In 1904 the first of these was unveiled in the churchyard of St Andrew's Church, just off Grafton Street. Dedicated to 'the gallant deeds of the 74th Dublin Company of the Imperial Yeomanry in the recent South African war', it was designed by the architect, A. E. Murray and unveiled by the Duke of Connaught on 6 May.[95]

> A guard of honour, with band and colour, furnished by the East Lancashire Regiment was drawn up outside the church in St Andrew Street, and presented a very fine appearance, all the men being of splendid physical proportions, and in every case displaying medals won in recent campaigns. A detachment of the south of Ireland Yeomanry with their band occupied a place inside the railings of the churchyard, and with them were many ex-members of the 74th co. The Southern Yeomanry detachment looked exceedingly smart, and their handsome green uniforms imparted a rich glow of colour to the scene.[96]

The Yeomanry memorial was followed by another dedicated to the officers and men of the five battalions of the Royal Dublin Fusiliers who were killed fighting on the British side in the Boer War. It was erected as an arched entrance gateway to St Stephen's Green in 1907.[97] Fusiliers' Arch was designed by John Howard Pentland together with Sir Thomas Drew and was formally dedicated by the Duke of Connaught.[98]

Memorialising a monarch: the statue of Queen Victoria, Leinster House

Large crowds assembled inside and outside the railings of the [Leinster] lawn some considerable time in advance . . . detachments from all the troops at present stationed in Dublin Garrison were in attendance . . . many spectators occupied positions on the balconies running round the Museum and the National Library, and from this elevated position obtained a splendid bird's-eye view of the entire proceedings . . . his Excellency pulled a cord . . . amidst an impressive roll of drums and the loud cheers of the spectators the figure of the Queen was revealed. The massed bands then played the national anthem, after which the drapery was removed from the other figures around the base of the monument.[99]

The idea of erecting a monument dedicated to Queen Victoria was originally mooted by members of the Royal Dublin Society (RDS) at the turn of the century. In April of 1900 a meeting of the council of the RDS was held at which a resolution was passed referring consideration of the desirability of opening a public subscription list for the proposed monument to Queen Victoria to a committee.[100] However, the project stalled and there was difficulty securing a site within the precincts of Leinster House owing to the belief that the monument should be erected by the public and not a private body (the RDS). Consequently, after the Royal visit of 1900 the committee appointed to look into the possibility of erecting a statue to the Queen met and proposed that 'the public at large should be invited to subscribe for the erection of the statue'.[101]

A public meeting was held in the lecture theatre of Leinster House on 8 May 1900:

> The proceedings throughout were of the most unanimous and enthusiastic character . . . it was proposed that the new subscription list should not be limited to any particular sum, and that the public at large should be invited to contribute. (Cheers)[102]

The discussions then focused on the site of the statue. The space between the National Library and the National Museum, in front of Leinster House, was considered most suitable, given its proximity to the statue of the Prince Consort which had been erected in Leinster Lawn by public subscription in 1872. The meeting concluded with the adoption of the resolution that:

> this meeting heartily approves of the project to erect in Dublin a statue of the Queen . . . that there was no statue of the Queen in Dublin, and that the citizens of the capital of Ireland felt that this state of affairs could not any longer be permitted to exist. It was the Queen and her personality, and not the throne and its splendour, that went home straight to the heart. They should have amongst them a statue of the Queen surrounded by her Irish soldiers.[103]

The Lord Chief Justice, in proposing this resolution overlooked the fact that nationalists referred to Queen Victoria as 'The Famine Queen', and it may have been more correct to note that it was in the hearts of the Anglo-Irish Ascendancy that she was held with such esteem.

Following the death of the Queen shortly after her visit to Ireland, the project gained renewed energy and work progressed swiftly. The sculptor John Hughes was approached and commissioned to make the monument for an inclusive fee of £7,000. He undertook to have it completed within five years.

The choice of contractors caused the committee considerable difficulty. They eventually selected the firm of Vienne of Paris which enjoyed 'a reputation second to none in France, and, of course, in that home of art there were enormous advantages at hand which should not be commanded here in Ireland'.[104] The committee also had protracted discussion regarding the choice of material to be used in the monument. It had originally been intended that the figure of the Queen would be cast in marble. However, Hughes ascertained from his contacts in France that marble would not withstand the Irish climate. As the *Weekly Irish Times* put it, 'if it is unsuitable for standing the weather conditions of the sunny land of France, it would be ruinous to expose such a work to the elements of this rainy country'.[105] It was decided, therefore, to cast the figure and the minor sculptures which would adorn the monument in bronze. In August 1907 work began on preparing the site to receive the monument and the laying of foundations commenced.

The monument was dominated by the bronze figure of Queen Victoria in a seated position, and with full regalia, surmounting a large pedestal and accompanied by a figure of Erin, who was depicted presenting a laurel wreath to a wounded Irish soldier (plate 3.15). On the other side of the monument, two bronze figures representing peace were incorporated and at the rear, facing Leinster House, a bronze figure of fame was positioned.[106]

> On Saturday afternoon the statue of the late Queen Victoria . . . was unveiled by the Lord Lieutenant, in the presence of a brilliant and distinguished gathering. The ceremony was made the occasion of an imposing military display . . . large crowds assembled inside and outside the railings of the lawn some considerable time in advance . . . detachments from all the troops at present stationed in Dublin Garrison were in attendance, and from two o'clock, when they began to arrive, Kildare Street and Molesworth Street . . . gradually assumed a highly congested appearance. About 1,000 troops were on parade, and as the various battalions, dressed in review order, with their bands, arrived on the scene, they took up their allotted positions with bayonets fixed, forming a wide circle round the greensward, in the centre of which stood the memorial, with the beauty of its magnificently formed bronze figures as yet unrevealed. The crimson covered passage leading from the entrance gates to the dais in front of the statue was lined by men of the Royal Scots Fusiliers, while across Kildare Street and along Molesworth Street the crowds were kept back by a long line of troops.[107]

Plate 3.15 **The unveiling of the Queen Victoria Monument at Leinster House, 1908**
From the *Capuchin Annual,* 1975

Saturday 15 February 1908 was the day selected for the unveiling of the memorial and the contemporary newspapers carried extensive coverage of the unveiling ceremony. The monument was formally handed over by Mr Justice Boyd on behalf of the memorial committee to Sir George Holmes, Chairman of the Board of Works amid a display of considerable pomp and ceremony, as *The Irish Times* reporter went on to note:

> Around the [Leinster] lawn were drawn up the Essex, the Warwick, the Royal Berkshire Regiments, and on the steps in front of Leinster House were the massed bands of those battalions . . . The whole scene, framed within the line of steel-tipped scarlet, with its magnificently proportioned architectural background, presented an exceedingly effective spectacle, and its beauty was enhanced by the generous sunshine, which agreeably tempered a sharp easterly wind.[108]

The vice-chairman of the subscription committee, Mr Justice Boyd, made a short speech in which he summarised events leading up to the erection of the statue and extolled the virtues of the late Queen to successive rounds of applause and cheering, before the Lord Lieutenant pulled the cord and unveiled 'amidst an impressive roll of drums and the loud cheers of the spectators the figure of the Queen'.[109] This was followed by the playing of the national anthem, after which the drapery was removed from the other figures around the base of the monument. The Lord Lieutenant then gave a short address and read a message from the King. Proceedings continued with speeches from other members of the committee, 'the national anthem was played as their excellencies left the grounds at the close of the ceremony; the troops returned to their quarters, and the public assembled within and without the gates dispersed'.[110]

It was not long, however, before criticism of the monument emerged. While the *Irish Builder and Engineer* echoed some of the sentiments expressed in *The Irish Times*, stating that, as a tribute of her Irish people and as a work of art, the monument was 'eminently satisfactory . . . the bronze figure of the Queen is dignified and imposing, while the character of the excellent base has a fine architectural appearance that accords satisfactorily with the surroundings and adequately fills in the centre of the fine square'.[111] It went on to comment that:

> the work was entirely executed in France. Possibly for this there may be justi-fication, because we all know, and must admit, that sculpture, alas! is not in a flourishing state in this country. The pedestal or base of the statue, is, however, the design of a French architect, and the work has been quarried, wrought, and carved in Vienne in France; for this we can see no justification. Ireland has some of the finest building stones in the world, and it is not inconceivable that a good design could have been procured and worked in some native material of proved durability. The history of foreign stone for outdoor work in this country is one of unqualified disaster . . . We cannot help fancying that this, being 'the tribute of her Irish people', that a work, the conception of Irish brains, and the work of Irish hands, would afford His Majesty the King more gratification than a work imported readymade into this country'[112]

After 1922 when Leinster House became the permanent location of the Irish Parliament (Dáil), the suitability of having a foreign monarch displayed in such a location was called into question. It was perhaps not surprising then that efforts began in earnest to have the statue removed.

Conclusion

The erection of each of these public monuments before 1922 served to create a representative landscape of some significance. The majority of those unveiled in the latter half of the nineteenth and the early twentieth century served to cultivate a nationalist monumental landscape in the heart of the Dublin city. Their unveiling, together with the not always successful insistence on using native, resident sculptors, can be interpreted as an attempt by various committees to challenge the ideology represented in the statues that had been erected in the eighteenth century and which had become outmoded amid the somewhat changed political and cultural context. Moreover, these sculptural initiatives also point to the increasing power and nationalist complexion of the city's governing authority, Dublin Corporation, consideration of which is essential in any assessment of the symbolic geography of Dublin before independence.

In the mid-nineteenth century, Dublin underwent a revolution in municipal government which paved the way for a new and reformed corporation. Prior to the passing of the Municipal Corporations Ireland Act in 1840 and the Dublin Improvement Act in 1849, the city's municipal authority was 'orange-dominated . . . the voice of Ascendancy and bigotry, operating under the principle of self-election and absolute control of admission to the franchise'.[113] The municipal reforms paved the way for the subsequent domination of the council chamber by nationalist-minded Roman Catholics at the expense of conservative, Unionist members. Throughout the remainder of the century many issues of little municipal significance but of great political and religious controversy were debated in its chamber.[114] The political composition of the corporation also served to alienate the Dublin Castle administration and from 1880 members of the authority boycotted official functions, although relations temporarily thawed with the arrival of the pro-Home Rule Viceroy, Lord Aberdeen. The nationalist agenda of Dublin Corporation ensured that it became firmly established 'as a body with wider political interests, a type of substitute for the lost Parliament of College Green'.[115]

This 'substitute Parliament' played a key role in granting permission for statues to be located on particular sites. Its willingness to sanction the erection of nationalist monuments in prominent locations was matched only by its reluctance to afford similar locations for statues that had a more loyalist bent. Much less conspicuous locations were granted to statues of, for example, the Earl of Carlisle, Lord Gough, Prince Albert and the Duke of Eglinton and Winton. This trend continued into the early decades of the twentieth century, most explicitly demonstrated when a monument to Queen Victoria was unveiled in the relative seclusion of Leinster House while a monument to Charles Stewart Parnell was unveiled at the head of Sackville Street.

The erection of the Sackville Street monuments in particular signalled the onset of a new era in the symbolic geography of Dublin. These statues stood alongside the pillar dedicated to Lord Nelson in an uneasy juxtaposition and served as a challenge in stone to the prevailing Castle administration and to the monumental landscape of imperial power that had been constructed centuries earlier. It is striking that these figures, drawn almost exclusively from Irish political, cultural and religious circles, should be unveiled in the heart of a city that remained a part of the British Empire. It would seem, however, that the Castle administration was impotent in attempting to hinder the creation of nationalist monumental landscape. Instead, from his offices in Dublin Castle, the chief secretary supervised the administration of a country that was politically and culturally deeply divided. Ireland may have been a colony of the British Empire, but its status as a colony was deeply ambivalent.

CHAPTER FOUR

Naming Dublin: imperial power, nationalist resistance

Interpreting the naming legacy

There are but few of our modern streets in this city called after distinguished national characters or events – indeed many of them are essentially English and foreign, and the few which have names 'racy of the soil' are are rather insignificant streets . . . the page of Irish history is brimful of names that deserve to be perpetuated in roads and streets.[1]

As urban space became an increasingly contested terrain in nineteenth-century Dublin, both nationalists and loyalists struggled to claim it as a political resource. Agencies loyal to empire sought to project an image of Ireland as an integral part of the United Kingdom, while nationalists by way of sharp contrast set out to construct an imaginative geography of Ireland as a separate nation that was both culturally and politically distinct. Processes of street naming and renaming therefore came to be particularly significant and highly contentious. Until the mid-nineteenth century the British authorities dominated the institutions of local government and consequently held considerable sway over the production of meaning in urban space. As the century progressed, however, and as the complexion of Dublin Corporation became increasingly nationalist, the balance of power shifted and naming became implicated in the symbolic reclamation of space by nationalists.

The names attached to the streets, bridges and quays in Dublin of the late nineteenth century derived from a broad range of sources (table 4.1). The city's medieval roots were recorded in a cluster of names to the west of O'Connell Street. These were drawn primarily from the many church buildings that were located throughout the medieval walled area and beyond, as well as from sources related to particular functions, hence Cook Street, Winetavern Street and Fishamble Street. During the eighteenth and nineteenth centuries, however, the capital became imbued with names that reflected the power of the Anglo-Irish ascendancy and served as tangible symbols of Ireland's place within the United Kingdom. Street names became an important means of celebrating collective memory, cultivating group identity and of creating a sense of a shared past and an official version of history.

Table 4.1 **Chief derivations of Dublin street names in the late nineteenth century**

	Origin
1	Kings and Queens
2	Lords Lieutenants
3	Noblemen
4	Lord Mayors
5	State officials
6	Property Owners
7	Celebrated persons
8	Public buildings
9	Churches
10	Taverns
11	Historic Places
12	Occupations
13	Titles of office
14	Physical features
15	Corruptions
16	Association of ideas
17	London names
18	Obsolescent names

Source: After C.T. M'Cready, *Dublin street names* (Dublin, 1892, reprinted 1975).

An analysis of the names in place in the 1880s reveals that Dublin's street names recalled past glories of the Empire, commemorated various members the British royal family, celebrated heroic British figures, and preserved the names of countless members of the Dublin Castle administration together with an array of nobles and wealthy landowners. Such names perpetuated in the streetscape and in the minds of inhabitants the memory of historical figures or events deemed worthy of remembering by those in charge. They were one component of a broader attempt to cultivate a sense of a shared past and to create an official version of history. Perhaps inevitably most of the street names honoured the priorities of the powerful rather than those who lived on the streets. The prominence afforded to the imperial iconography, and the corresponding absence of a nationalist iconography, identified this space as territory of empire. Such names also reflected the centrality of the monarchy to unionist identity, a fact that was reinforced in the monuments erected in the capital.

The names assigned to the city's quays and bridges in the eighteenth and early nineteenth century at a time when the ruling elite had strong connections with the colonial power provide a useful framework for exploring the legacy of street nomenclature (figure 4.1). These names placed clear emphasis upon Britain's national history, its military heroes, illustrious monarchy and imperial achievements (see tables 4.2, 4.3, 4.4 and 4.5). Moreover, they contributed to the cultivation of an official version of Irish history which placed Ireland firmly

Table 4.2 **The names and derivations of Dublin's bridges in 1922**

Name	Probable Date	Derivation
Sarah	1791	After Sarah, Countess of Westmoreland, wife of the Lord Lieutenant, 1790–5
King's	1828	After visit of King George IV in 1821
Victoria & Albert	1863	After Queen Victoria and Prince Albert who used the bridge during a royal visit in 1861
Queen's	1776	After Queen Charlotte, the wife of George III
Whitworth	1818	After the Earl of Whitworth, Lord Lieutenant, 1813–17
Richmond	1816	After the Duchess of Richmond who laid the foundation stone
Grattan	1875	Formerly Essex Bridge after Arthur Capel the Earl of Essex, Lord Lieutenant, 1672–7. Rebuilt in 1755 and again in 1875 when it was renamed after Henry Grattan, MP
Wellington	1816	After the Duke of Wellington who won the Battle of Waterloo in 1815
O'Connell	1880	Formerly Carlisle Bridge after the 5th Earl of Carlisle, the bridge was renamed after Daniel O'Connell, MP
Butt	1879	After Isaac Butt, MP and QC

Sources: Compiled using C.T. M'Cready, *Dublin street names* (Dublin, 1892) and J. de Courcy, *The Liffey in Dublin* (Dublin, 1996)

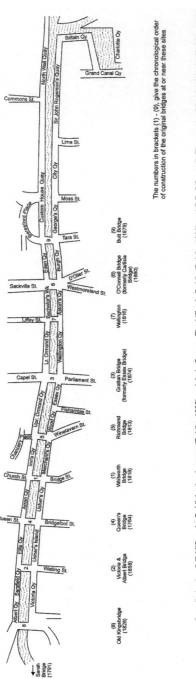

The numbers in brackets (1) - (9), give the chronological order of construction of the original bridges at or near these sites

Compiled using C.T. M'Cready, *Dublin's street names* (Dublin, 1892) and J. de Courcy, *The Liffey in Dublin* (Dublin, 1996) and E.E. O'Donnell, *The annals of Dublin, fair city* (Dublin, 1987)

Figure 4.1 **The names of Dublin's quays and bridges in 1922**

Table 4.3 **The names and derivations of the Grand Canal Docks quays in 1922**

Name	Probable Date	Derivation
Great Britain	1795	After Great Britain
Hanover	1800	After King George I, 2nd elector of Hanover, 1714–27
Grand Canal	1791	From the Grand Canal opened here in 1791
Charlotte	1795	After Queen Charlotte, wife of King George III
South Wall	1766	From being south of the Liffey

Sources: Compiled using C.T. M'Cready, *Dublin street names*, (Dublin, 1892) and J. de Courcy, *The Liffey in Dublin* (Dublin, 1996)

Table 4.4 **The names and derivations of the north quays in 1922**

Name	Probable Date	Derivation
Albert	1871	After the Prince Consort
Sarsfield	1885	Formerly Pembroke Quay, after Thomas Herbert, 8th Earl of Pembroke, Lord Lieutenant 1707–9, renamed after the patriot, Patrick Sarsfield
Ellis	1766	After Sir John and William Ellis, property owners
Arran	1728	After Lord Arran, son of James, Duke of Ormond, and his deputy in Ireland from 1682 to 1684
King's Inn	1743	From the Inns of Court which stood on the site now occupied by the Four Courts
Ormond	1728	After James Butler, 1st Duke of Ormond, Lord Lieutenant 1643–9, 1660–9
Bachelors Walk	1728	Uncertain origin, thought to be after a property owner
Eden	1796	After the Rt. Hon., William Eden, Chief Secretary 1780–2
Custom House	1791	After the Custom House, opened in 1791
North Wall	1756	From being north of the Liffey

Sources: Compiled using C.T. M'Cready, *Dublin street names* (Dublin, 1892) and J. de Courcy, *The Liffey in Dublin* (Dublin, 1996)

Table 4.5 **The names and derivations of the south quays in 1922**

Name	Probable Date	Derivation
Victoria	1871	After Queen Victoria
Usher's Island	1562	After an island formed by a branch of the river Camac which divided at the northern end of Watling Street. Site of a Mr Usher's House
Usher	1728	From proximity to the house of Usher
Merchant	1565	From its being occupied by merchants
Wood	1451	From the fact that the quay was formerly made of wood
Essex	1756	After Arthur Capel, Earl of Essex, Lord Lieutenant, 1672–7
Wellington	1817	After Arthur Wellesley, Duke of Wellington
Crampton	1766	After Philip Crampton, Lord Mayor 1758, a wealthy bookseller and property owner
Aston	1708	After Henry Aston, merchant
Burgh	1808	After Elizabeth Burgh, wife of Anthony Foster, Chief Baron of the Exchequer
George	1723	After King George I
City	1766	After the city
Sir John Rogerson	1728	After Sir John Rogerson, Lord Mayor, 1693, MP for Dublin, 1695, property owner

Sources: Compiled using C.T. M'Cready, *Dublin street names*, (Dublin, 1892) and J. de Courcy, *The Liffey in Dublin* (Dublin, 1996)

within the United Kingdom, while also marginalising unofficial and competing narratives and political traditions. By the later decades of the nineteenth century, however, this began to change as the power of the Anglo-Irish elite continued to wane and local government reforms enabled nationalists to assume power and challenge the official narratives that had been symbolically woven into the landscape. Street naming and more significantly renaming became an important tool in asserting a sense of Irish national identity. As nationalists assumed control over the institutions of local government, they inherited a symbolic landscape that celebrated and legitimated the union between Ireland and Britain. But as they exercised their authority over the urban landscape through the renaming of streets *before* the achievement of political independence, the ascendant hegemony and its vision of Ireland was challenged, critiqued and deconstructed in the domain of the urban spaces that it had itself appropriated.

Renaming streets: demonstrating resistance

Towards the close of the nineteenth century the dissatisfaction held by many with the street nomenclature of the capital city became increasingly vocal. As a columnist in *The Irishman* remarked in 1862:

> A thorough reform of the nomenclature of Dublin street names [is called for] . . . the names of some of our streets, lanes and places are barbarous – and most of them foreign and anti-national . . . Grafton Street, for instance should be Grattan Street; Nassau Street should be abolished, and merged in Leinster Street, Britain Street should be Burke Street and Brunswick Street should be Brian Street. Then we should have Sheridan Street, Moore Square, Goldsmith Place . . . The beginning must be made by changing Sackville Street to O'Connell Boulevard. Unless it is intended to dishonour O'Connell this must be done.[2]

Indeed, in an address that he made in 1892, the future Irish President, Douglas Hyde, looked forward to the emergence of a united and independent Irish government that would set in place changes to the street nomenclature of the city. He observed that:

> On the whole, our place names have been treated with about the same respect as if they were the names of a savage tribe which had never before been reduced to writing, and with about the same intelligence and contempt as vulgar English squatters treat the topographical nomenclature of the Red Indians. These things are now to a certain extent stereotyped, and are difficult at this hour to change . . . it would take the strength and goodwill of an united nation to put our topographical nomenclature on a rational basis like that of Wales and the Scotch Highlands . . . I hope and trust that where it may be done without any great inconvenience a native Irish government will be induced to provide for the restoration of our place names on something like a rational basis.[3]

The fact that street names could serve as powerful tools in the expression of dissent and opposition to the established regime and could anchor alternative narratives of identity was recognised in the chamber of Dublin Corporation. In the latter decades of the nineteenth century a raft of proposals to rename streets in the city were put forward. This trend demonstrated that renaming streets, just like erecting public monuments, had an immediate effect on the politics and symbolism of landscape and served as an act of political propaganda of enormous symbolic value. The renaming of one of the city's most important bridges in 1880, that linked Dublin's central thoroughfare Sackville Street to the south side of the city, from Carlisle Bridge (after the 5th Earl of Carlisle) to O'Connell Bridge (after Daniel O'Connell, a nationalist MP) heralded the arrival of a new naming era. From then on Dublin Corporation set about renaming streets in an effort to install a nomenclature that was more in keeping with their nationalist political agenda.

In 1884 the Corporation passed a significant motion which called on a committee of the whole house 'to take into consideration the street nomenclature of the city, with the view to changing the names of some of the prominent thoroughfares'.[4] The committee reported back in 1884:

> Your committee have taken into consideration the foregoing Order of Reference, and beg to report that they consider the names of the main thoroughfares of the chief cities of a nation should be such as to recall events in its history and progress deserving of commemoration; and that the names of many of the principal streets in Dublin do not do so, several of these names being without meaning to the present resident, the streets possibly having been called after insignificant persons undistinguished by any public services.[5]

It went on to suggest that any changes should necessitate the least possible inconvenience to commerce and the postal service. Consequently, the only change that it recommended was that Sackville Street should become O'Connell Street, 'a name which will ever be connected with the greatest moral and social reforms of this century and nation, and one locally associated with the existing magnificent national monument and noble bridge'.[6]

From Sackville to O'Connell Street

The issue of renaming Sackville Street, which had been named after a former Lord Lieutenant, Lionel Cranfield Sackville, the first Duke of Dorset and Lord Lieutenant 1731–7 and 1751–5, to O'Connell Street, after Daniel O'Connell, whose statue had been unveiled in 1882, evoked a mixed reaction. After the publication of the 1884 report, a deputation of inhabitants from Upper and Lower Sackville Street attended the council meeting on 10 October. They 'requested the Corporation not to make the change, and presented documents in support of their views, signed by the residents'.[7] The Dublin Chamber of Commerce added its voice to the debate when they submitted to the

Corporation a copy of a resolution which was unanimously passed at one of their meetings.

> It is the opinion of this council that the project of renaming the streets of Dublin would, if carried into effect, be most injurious to the interest of the trading and commercial classes and to the community in general, and that it would create a vast amount of commercial and postal confusion and inconvenience, and that a copy of this resolution be sent to the Lord Mayor.[8]

The report of the committee of the whole house came up for consideration in December of 1884, although it was postponed for one week in order to allow petitions from the residents of Upper and Lower Sackville Street opposing the name change to be considered.[9] The critical meeting took place on 8 December and was attended by 'a deputation of owners and occupiers of houses in Upper and Lower Sackville Street . . . A. D. Kennedy, Esq., Paul Askin, Esq., Alex Findlater, Esq., and Samuel M'Comas, Esq., addressed the Council on behalf of the deputation, and urged the Corporation not to change the name of Sackville-street, as they considered such change would be a direct interference with the rights of property, cause confusion in the titles to houses, and be detrimental to their business.'[10] An amendment was put that 'in deference to the wishes of the owners and occupiers of the houses in Sackville Street, the recommendations contained in the Report . . . be not adopted'.[11] This amendment was voted down by a vote of 12–31 and Dublin Corporation formally ratified the change.

While the Corporation may have wished to change the name of the street, the residents and traders were not particularly enthused by the proposal and rejected it outright. They took the matter to the courts the following year. In January 1885 a copy of a writ seeking an injunction restraining Dublin Corporation from making any change was served on the town clerk at the suit of Joseph Anderson and other Sackville Street traders.[12] They argued that the corporation lacked the authority to make any such change and that the vast majority of residents were opposed to the proposal. They also stated that the change would cause inconvenience and injury to trade. While Dublin Corporation once again voted in favour of the name change,[13] the vice-chancellor, Hedges Eyre Chatterton, whose task was to adjudicate on the matter, found in favour of the residents, and granted the injunction. He pointed out that the Corporation did not in fact possess the power to make name changes and in his report stated that:

The reason assigned . . . for the proposed change in the name, not only of
Sackville Street, but also of many others of the principal streets, is that [Dublin
Corporation] consider that the names of the main thoroughfares of the chief
cities of a nation should be such as to recall events in its history and progress
deserving of commemoration, and that the names of many of the principal
streets in Dublin do not do so. This may be a laudable notion in the abstract,
but, certainly, such sentiment and historical grounds are not those upon which
a municipal body can lawfully or reasonably act in the exercise of its statutory
functions. If the proposed exercise of these functions will be productive of
material injury to the property in the locality, and of serious inconvenience to
the persons immediately connected with it, there is no justification for such
action to be found in these sentimental notions . . . it must be borne in mind
that in matters of sentiment there is no certainty or even reasonable probability,
of continuance; and that these notions which may be now prevalent may – and
probably will – in a few years be changed or exploded, and so on from time to
time. If old and established names be now changed on such grounds, the
probability is that other changes will follow, and possibly be frequent. It is plain
that a state of uncertainty must thereafter arise, highly injurious to property,
and productive of great personal inconvenience and of pecuniary loss which
may also be considerable.[14]

In the wake of the granting of the injunction, Dublin Corporation set
about putting in place the legal mechanisms to ensure that they could affect
changes to street names. The Dublin Corporation Act of 1890 provided clear
authority for the Corporation to alter the names of the City's streets. With
regard to Sackville Street, however, although the street had begun to be known
unofficially by the name of O'Connell Street, it was not until after the achievement of political independence that the authority formally brought the name
change into effect.

Naming and renaming in late nineteenth-century Dublin
Although it was not until after independence that Sackville Street was formally
renamed O'Connell Street, in the meantime a number of other name changes
were proceeded with and new names assigned (table 4.6 and figure 4.2).[15] The
newly laid-out street from Great Brunswick Street to Butt Bridge was named
Tara Street after the ancient meeting place of the Irish High Kings.[16] Soon after,
in 1885, the route-way from Cork Hill to Christchurch Place was named Lord
Edward Street, after the Irish patriot, Lord Edward Fitzgerald,[17] while the

proposal was carried to change Pembroke Quay to Sarsfield Quay, after Patrick Sarsfield.[18] The years through to 1922 were also marked by a set of proposals to change a range of other street names, and not always for overtly political reasons. In 1885 for example, Lower Temple Street was renamed Hill Street.[19] This change was the subject of a report that was read at a meeting of the Corporation in November 1885.

> Your committee for some time past have had under their consideration memorials from the inhabitants and owners of house property in Upper Temple-street, praying that the name of the latter street may be changed, in consequence of the very serious deterioration in the value of property in those streets, owing to the fact that certain houses in Lower Temple street have been for some time past occupied by immoral characters. And taking into consideration all the circumstances of the case, as set out in the memorials above referred to, your Committee beg to recommend in the strongest manner that the prayer of those memorials be granted, and that the name of 'Lower Temple-street' be changed to that of 'Hill-street,' in sufficient time to permit of the change being duly recorded in Thom's Directory for the year 1886. All of which we beg to submit as our Report for this 13th day of October, 1885.[20]

The Corporation also received numerous memorials which alluded to the 'disreputable character that had for some years past been attached to Lower Temple-street, in consequence of the immoral houses located there'.[21] It was suggested that 'as the street was built on a steep hill, the name of Hill Street would be a suitable one'.[22] Other proposals included the renaming of Upper Mecklenburgh Street to Tyrone Street.[23] This proposition owed more to local snobbery than to politics. The report issued regarding the change noted that:

> Your committee have had under consideration a numerously-signed memorial (copy of which is appended hereto) from owners of property and residents in Upper Mecklenburgh-street, and immediate neighbourhood, praying that the name of that street may be changed, and giving as their reasons for making the application that the inhabitants of Upper Mecklenburgh-street are members of the respectable working classes, while many of the houses in Lower Mecklenburgh-street are used for improper purposes, and inhabited by persons of the worse character. Your committee being strongly impressed by the reason given in the memorial in question, are decidedly of opinion that the changing of the name of Upper Mecklenburgh-street would be most desirable, and,

under the circumstances, absolutely necessary, and therefore beg respectfully to recommend the Municipal Council to sanction that street in future being called Tyrone-street. By this arrangement Lower Mecklenburgh-street will be the only street bearing that name, which for a long time past has been a notorious one.[24]

The report was adopted and in 1888 the issue of renaming Lower Mecklenburgh Street was considered.[25] Residents of the street, 'respectable working-class people', proposed that it 'be renamed Drogheda-street'.[26] The residents put forward much the same arguments as had been used by their neighbours some years previously, that 'the improvement of that locality is incomplete so long as the name of "Mecklenburgh" attaches to any portion of that street',[27] and suggested that the street be renamed Lower Tyrone-street, or Tyrconnell, Armagh or Drogheda street:

> In addition to the injury done to valuable property in the neighbourhood by the immediate vicinity of Mecklenburgh-street, there are in the street itself a number of humble residents with grown families, the female members of which find it impossible to obtain employment as domestic servants owing to their residence in a locality to the name of which such bad odour attaches.[28]

The change was eventually made later in 1888 when it was renamed Tyrone Street.[29]

By 1900 the text of Dublin's street names had come to reflect the increasingly contested nature of space in the city. The inherited legacy of previous generations pointed to the connections with the British Empire, which did not, however, sit easily with the nationalist complexion of Dublin Corporation. It was during the years from 1900 through to 1916, and more so in the immediate aftermath of the achievement of independence, that the renaming of streets was used as a counter tool by the municipal authority to challenge the dominance of the inherited legacy. Moreover, the issue of bilingual naming and the role of the Irish language gathered pace in the early years of the twentieth century. Dublin Corporation passed a number of further street name changes, a central strain of which involved the 'sanitisation' of the north inner city. For example, Lower Gloucester Street was changed to Killarney Street because of its 'very undesirable reputation of late years, in consequence of which the houses in Upper Gloucester Street have deteriorated in value . . . most, if not all of them, are set in tenements. This state of things has also had the effect of considerably depreciating the value of the houses on Lower Gloucester Street.'[30] In 1906 Montgomery Street was renamed Foley Street, after John Henry Foley the sculptor of many of Dublin's late nineteenth-century monuments.[31] Another street in this area that was 'sanitised' by name change was Mabbot Street. It became Corporation Street in 1911, while Tyrone Streets Upper and Lower were renamed Railway Street and Waterford Street respectively in the same year.[32] There were also more politically overt motives behind the street renaming strategy, which is evident in the renaming of Great Britain Street.[33] This became known as Parnell Street in 1911, the same year that the statue of Parnell was erected at the northern end of O'Connell Street, when it was moved that:

> The name of Great Britain Street be and it is hereby changed to Parnell Street, in recognition of the great service rendered to Ireland by the late Charles Stewart Parnell. The Rt. Hon. the Lord Mayor with the leave of the Council added to his motion the following words: 'The majority in number and value of the ratepayers in the street having expressed their consent to the change'.[34]

The Irish language and street names

Another dimension important to consider when exploring the city's street names is that of the language in which these names were represented. Before 1900 names appeared in the English language. Over the course of subsequent years, however, a series of motions were introduced which sought to give greater prominence to the Irish language. In 1900 a motion was moved in the council chamber that 'it be an instruction to the paving committee to have erected and conspicuously displayed tablets bearing the names of the streets, lanes, squares, etc., of this metropolis printed in Gaelic.'[35] Although the motion was voted down by 17 to eight votes, it was a signal of future developments.[36] In 1900 the Paving Committee of Dublin Corporation issued a report 'relative to the erection of bilingual street nameplates', in which it was stated that:

> The question of the erection of these nameplates has been brought prominently under our notice by various associations interested in the revival of the Irish language. The question of the cost involved in the scheme being one of great importance when considering as to what action we should take in the matter, we instituted inquiries, and as will be seen from the Borough Surveyor's report, the cost of erecting bilingual street name-plates at 8/- each – the ascertained price – in the different streets of the city, including those within the

Table 4.6 **Streets renamed or proposed for renaming, 1880–1922**

Original name	Proposed change	Year	
Gregg's Lane	Findlater Place	1882	
Sackville Street	O'Connell Street	1884	1922
Pembroke Quay	Sarsfield Quay	1885	
Lower Temple Street	Hill Street	1885	
O'Connell St (Berkeley Rd)	O'Connell Avenue	1886	
Bow Lane	Swift Street	Withdrawn	1886
Sackville Lane	O'Connell Lane	Voted down	1886
Sackville Place	O'Connell Place		
	Eblana Lane/Place	Voted down in 1886	
Upr Mecklenburgh Street	Tyrone Street	1886	
Walker's Alley	St Patrick's Place	1886	
Lr Mecklenburgh Street	Lr Tyrone Street	1888	
Barrack Street	Benburb Street	1889	
Tighe Street	Benburb Street	1889	
Usher's Quay to New Row	St Augustine Street	1889	
Pill Lane	Chancery Street	1896	
Love Lane	Donore Avenue	1896	
Lower Gloucester Street	Killarney Street	1901	
Montgomery Street	Foley Street	1907	
Mabbot Street	Corporation Street	1911	
Great Britain Street	Parnell Street	1911	
Tyrone Street Lr	Railway Street	1911	
Tyrone Street Upr	Waterford Street	1911	

Sources: Compiled using Minutes and RPDCD, 1880–1916

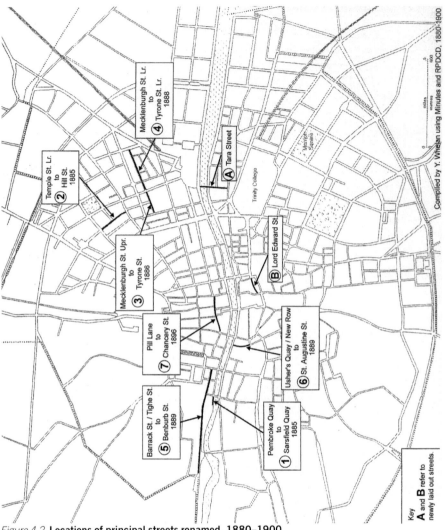

Figure 4.2 **Locations of principal streets renamed, 1880–1900**

added area, will amount to a sum of £3,400, for which, of course, there is no provision in our finances. We have, therefore, thought it advisable, considering the large amount involved in the erection of such name-plates, to bring the whole question under the consideration of the Municipal Council, in order that they may determine what action we should take in reference to this matter.[37]

It was subsequently passed by the Corporation that 'in the cases where street plates are renewed or new ones put up they should be lettered bilingually'.[38] This effectively ruled out the large-scale erection of bilingual nameplates.

In 1904 the Irish language issue found expression in the notice of a motion by Alderman Cole:

That this Council, desiring to meet the widely-expressed wishes of so many thousands of the citizens in the Irish as well as in the English language, direct that estimates be advertised for showing at what cost, and in what Irish material the names of the streets throughout the city could be affixed to the electric lamp poles, and that a special sub-committee of this council be appointed to draw up a list of names in Irish to be so placed – the present names of the streets in English to remain as they are. [39]

The motion was ruled out of order, however, on the grounds that the council had already considered this matter and decided that when new nameplates were to be affixed they should be bilingual.[40] In 1909 it was moved:

That the paving committee be directed to include in their next Breviate a list of the streets, roads, lands, courts and other places where bilingual name-plates have been erected, such list to give the Irish and English form of each name; and that a revised list of such names in Irish and English be in future included in the Breviate coming before Council at the end of each financial year until the erection of bilingual nameplates throughout the entire city has been completed.[41]

The motion was passed and in 1910 it was moved that for the information of members and officers of the Council, the Town Clerk, acting in conjunction with the Paving Committee:

be directed and empowered to have prepared and to issue a list, in English and Irish, of all streets, roads, avenues, and other such places in the city of Dublin. That the town clerk and secretary of the Paving Committee shall arrange to hold an interview at the City Hall with the secretary of the Gaelic League, and J. H. Lloyd, who is a well known authority on such matters, for the purposes of considering the best means of carrying this into effect. This resolution shall not apply to any obsolete or unnecessary names of rows, villas, or terraces, forming part of a street or road, where the numbers now run consecutively.[42]

Conclusion

As the city stood on the threshold of a new era, Dublin's symbolic geography had come to embody the struggle for superiority, victory and ultimately power, that persisted between Britain and one of its kingdoms, Ireland. This struggle left turn of the century Dublin in something of a schizophrenic position. On the one hand it was a city of the British Empire and capital of the Kingdom of Ireland within the confines of the United Kingdom of Great Britain and Ireland, and the canvas upon which the British administration set out to paint a picture of union and loyalty. On the other it was a city under the local governance of the strongly nationalist Dublin Corporation which attempted to assert a tangible sense of Irish national identity upon the urban landscape in the years before 1922.

The issues that were raised over the course of the decades from 1880 to 1921 highlight the powerful nature of the role that nomenclature has to play in the symbolic geography of place. In Dublin, three issues predominated: the eradication of the colonial legacy, the sanitising of certain sectors of the city, especially the north inner city, and the role of the Irish language on street nameplates. The commemorative street names in place in the capital served to conflate history and geography in a process that produced an official, authorised version of the past and represented an iconographic layer of imperial power that placed Dublin and Ireland firmly within the confines of the United Kingdom and the broader British Empire. This was increasingly contested towards the close of the century, however, when the politically motivated renaming of streets proved to be a highly demonstrative act of significant symbolic value. They served to cultivate an alternative political narrative and what is particularly striking is that these name changes occurred before the achievement of political independence in Ireland. After 1922 the practice became an even more important 'ritual of revolution' that contributed to the nation-building agenda of the Free State administration.

This contested political context was also reflected in the public architectural initiatives that took place in the latter decades of the nineteenth century. Once Ireland had become firmly incorporated into the British Empire, a significant cluster of cultural institutions were built in the south inner city in the vicinity of Leinster House. Chief among these was the National Gallery, Natural History Museum, National Library and National Museum. These were intended as sister institutions to those built in Edinburgh or London and in terms of the capital's symbolic geography reinforced the city's place within the broader empire. This was echoed in the early twentieth century with the construction of the last architectural project of the British administration in Ireland, the Royal College of Science, also on Merrion Street. As the British Empire spread its influence across the globe during the reign of Queen Victoria, projects of architectural grandeur, alongside public statuary and street names served as statements of a shared British identity. Such statements were, however, to be forcefully challenged in the early decades of the twentieth century and especially after the establishment of the Irish Free State in 1922.

PART III

A capital once again:
reinventing Dublin after independence

Plate 5.1 **'The last hour of the night'**
From P. Abercrombie et al., *Dublin of the future* (Dublin, 1922), frontispiece by Harry Clarke

CHAPTER FIVE

Symbolising the state: planning, power and the architecture of nationhood

The dawn of an independent era

The great thoroughfare which the citizen of Dublin was accustomed to describe proudly as 'the finest street in Europe' has been reduced to a smoking reproduction of the ruin wrought at Ypres by the mercilessness of the Hun. Elsewhere throughout the city streets have been devastated, centres of thriving industry have been placed in peril or ruined, a paralysis of work and commerce has been imposed, and the public confidence, that is the life of trade and employment, has received a staggering blow from which it will take almost a generation to recover.[1]

The Easter Rising began in April 1916, signalling the onset of several years of destruction, death and civic unrest, punctuated by the War of Independence and culminating in the Civil War.[2] During these years sectors of Dublin were destroyed, in particular the area around Sackville Street, as well as some of the city's chief public buildings, notably the General Post Office, the Custom House and the Four Courts (plate 5.1).[3] The rebellion raged for one week and saw large portions of the city centre reduced to rubble, while a number of strategically located buildings throughout the inner city were occupied.[4] By the following Sunday the city was quiet. Four hundred and fifty people had been killed, 350 of them civilians, while a further 2,614 were wounded and damage was estimated at £2 million. The focal point of British communications in Ireland, the General Post Office, was crippled and the building left in ruins. It was not the only one. As the annual report of the Commissioners of Public Works (OPW) noted for 1917:

During the disturbances in the month of April a number of public buildings suffered injury. The GPO was almost wholly destroyed, little more than the outer walls escaping. The extensive premises known as 'The Linenhall Barracks' were extensively destroyed; Coleraine House, a separate building, alone escaped injury. Considerable damage was also done to parts of the Four Courts buildings, which has since been made good. Following the destruction of the General Post Office temporary arrangements were made for providing accommodation for the various staffs of the department to enable postal and telegraph business to be transacted.[5]

The initial destruction and tentative reconstruction which occurred in the aftermath of the 1916 Rising was overshadowed by more destruction during the War of Independence.[6] It began in September 1919 and raged until July 1921, during which time more than 700 people were killed and 1,200 wounded. Houses and shops were extensively damaged, along with strategically and symbolically significant public buildings, such as the Custom House. It came under attack by the Dublin Brigade of the IRA in May 1921. In the ensuing battle with British forces the square in front of the Custom House was left, as contemporary news reports describe, 'full of pools of blood and people, dead and wounded, lay huddled in all directions'.[7] As the *Irish Builder* reported:

> The fire was caused by the act of a band of armed men, who at midday during the luncheon hours entered the building, bringing with them petrol and other material, and proceeded with the utmost coolness, to set fire to the building, having first warned and collected together the staff therein. In the conflict that ensued between Republican and Crown forces, unhappily eight lives were lost. How thoroughly the raiders performed their task, the burnt-out ruins tell. . . . It is just over five years since we had to mourn the destruction of another of the capital's great buildings, the General Post Office, one of that noble series of edifices which helped to raise eighteenth century Dublin to a foremost place amongst the capitals of the world, but still a building of far less artistic value than the custom house . . . Of the many fine buildings of Dublin, the Custom House was incomparably the best.[8]

In the wake of the exchanges fire spread through the building and efforts to save it or any of the contents were in vain, such was the intensity of the blaze which continued to smoulder for weeks. The destruction of the building ensured that one of the most important branches of British civil government in Ireland was reduced to virtual impotence, although the British government maintained that 'The burning of the Custom House causes more inconvenience to the public than to Government departments.'[9]

While a truce was brokered in July 1921, the Anglo-Irish Treaty that was signed in December 1921 led to a split in republican ranks which erupted into civil war in June 1922.[10] In the capital, the war that raged between the pro-Treaty government forces and the anti-Treaty irregulars, led to the occupation once more of key buildings in the capital by the irregular forces. Among them lay the Four Courts. It was occupied by the irregulars under the leadership of Rory O'Connor who used it as their headquarters until the building was bombarded on Wednesday, 28 June 1922 by the government troops.[11] The refusal of the irregulars to surrender led to the further bombardment of the building by rifle, machine gun and artillery fire and by the evening of the following day it lay severely damaged, 'breached in many places and fire throughout fuelled by the mass of timber which had been used in its construction. The government troops gained possession of the building and the irregulars surrendered shortly afterwards.'[12]

After the Four Courts had fallen, the troops of the Free State turned their attention to Upper Sackville Street, which was also heavily occupied by irregular forces. The upper part of the street was subsequently razed, echoing the destruction of Lower Sackville Street which had occurred five years earlier:

> The block of buildings in Upper Sackville Street, stretching from the Tramway Office to Findlater Place was also in the occupation of the irregulars, and after the fall of the Four Courts, the government troops turned their attention to ejecting their opponents from Sackville Street. The result of this movement is that the most important part of the street now lies in ruins.[13]

Thus, by July 1922, 'three of the great buildings that impressed the stranger with the conviction that this was indeed a metropolis are in ruins'.[14]

The sheer extent of the destruction, coupled with the stagnant economic conditions inherited by the Free State, highlighted the need for both political stability and the formulation of economic development strategies for the independent state. While the country's new leaders set about the business of political and economic development, various aspects of the urban cultural landscape came to play a significant role in marking the transition from dependence to independence.[15] This was expressed in a range of ways. Post-boxes were painted green covering over although not obliterating the Royal insignia of *Victoria*

Regina or *Edwardus Rex*, while the new State Seal featured an Irish Harp that was reputed to have belonged to Brian Boru who had been killed at the Battle of Clontarf in 1014. In later years a set of national stamps was issued and the national coinage was launched along with the State passport.[16] The Irish language also had an important role to play when, during de Valera's administration (1932–48), Irish titles were used for State agencies in an effort to confront citizens with the visual presence of the language.

In the capital city, meanwhile, attempts were made to rename streets, public monuments were erected *and* destroyed, and new planning and architectural initiatives were drawn up to give expression to the city's new-found status as capital of the independent Free State. Dublin Corporation, for example, played a key role in attempting to change the street names in the city, while central government was instrumental in establishing a site for parliamentary democracy. A range of *ad hoc* committees put forward proposals for commemorative monuments and subversive groups used illegal methods to obliterate the city's imperial iconography. This chapter begins by examining the issues and debates that surrounded the proposed building of a parliamentary complex for the new capital. Although this venture never came to pass, it nevertheless sheds light on the role of public buildings, and more specifically the parliamentary complex, in contributing to the inscription of national identity in the cultural landscape. The emphasis then shifts in subsequent chapters to an exploration of the role of urban planning initiatives, street naming and renaming and public monuments in the nation-building process. As Ireland entered a new era these aspects of the cultural landscape took on heightened significance just as it became, 'a national duty to ensure that the national capital shall worthily reflect the spirit of a free Ireland'.[17]

Building a capitol complex for a capital city

Dublin . . . it is the national capital, the legislative, the administrative, the traffic, the commercial, the professional, the cultural, and the social headquarters not alone of the new Irish State – but, in a very real sense, of the greater Ireland beyond the seas. It will of necessity be the highest expression of our national being – the focus of our national effort. What an opportunity! What a responsibility![18]

The years of civic unrest had had profound implications for the Dublin cityscape and not least its architectural fabric. Many of the city's most prominent public buildings, which had been focal points of the British administration, lay in ruins, notably the Custom House, Four Courts and General Post Office, along with much of the city's central thoroughfare, Sackville Street. One of the questions which confronted many in the years immediately following 1922, was whether or not these buildings, which had been constructed during the era of British rule, should be entirely demolished or reconstructed.[19] Among the many pressing issues that faced the government was that of where to house the seat of government. As one of the chief symbolic markers of independence, the parliament building is much more than a functional entity. Rather, the placement of government headquarters is an exercise in political power, a spatial declaration of political control and it is invariably a product of financial, political and cultural forces. As Vale writes, 'the decision to build a new place for government is always a significant one; the decisions about where and how to house it are more telling still'.[20] Fledgling states are often faced with two choices – to build a new parliamentary complex (or even a new capital city) with all of the symbolic potency that that entails, or to re-use existing buildings, which may have been associated with the previous regime. This very choice faced Ireland's leaders after the establishment of the Free State when debate raged over whether Leinster House, the headquarters of the Royal Dublin Society (RDS) and temporary location of the government since 1922, would continue to be used or whether a new site for a capitol complex would be found.[21] The debate was further complicated by the existence on College Green of the former Parliament House. Once the site of Grattan's Parliament in the eighteenth century, many argued it should revert back to its former, pre-Act of Union usage in a symbolic reclamation of space. One contemporary commentator alluded to the nature of the dilemma:

> A national parliament house and group of modern Government buildings is urgently required. Is the already overcrowded College Green banking and commercial centre to be invaded, or is universal modern practice to be followed, and our new legislative assembly to be permitted to deliberate in suburban surroundings which are detached from the atmosphere and distractions of commercial activity?'[22]

Between September 1922 and February 1924 a protracted debate raged regarding the location of the proposed new national parliament building. The temporary location of government in Leinster House had forced, for security reasons, the closure of the National Gallery, Library and Museum and also necessitated the relocation of the RDS to more cramped conditions.[23] The issue became even more pressing owing to the pressure on space which the destruction of the key public buildings had created.[24] As the OPW annual reports reveal, 'The arrangements for occupation by the *Oireachtas* of temporary premises were not found very satisfactory. Neither the *Dáil* nor the *Seanad* was well pleased with the accommodation provided, and the exclusion of the public from the Museum, and the exclusion of the members of the Dublin Society from a part of their premises gave rise to complaint.'[25] A range of proposals were subsequently put forward, among them those of the Greater Dublin Reconstruction Movement (GDRM), a group comprising leading members of the town planning movement in Ireland and led by E.A. Aston.

The GDRM and plans for a parliamentary complex

The GDRM was formed in 1922 under the chairmanship of Senator James Moran and was led by key members of the town planning movement in Ireland – E. A. Aston, W. Purcell O'Neill, Lady Aberdeen and F. Mears. Aston was the major motivating force and he announced their principal proposals in a lecture to the Engineering and Scientific Association of Ireland in December of 1922. This was later published as the *Greater Dublin reconstruction scheme described and illustrated* and was exhibited at the Mansion House, 1–12 January 1923.[26] In his address Aston outlined the main proposals and recommendations of the movement, structuring his lecture under a number of headings (table 5.1). With regard to the capital's destroyed public buildings, the GDRM made a number of radical suggestions which implied the wholesale reworking of the geography of these buildings (plate 5.2). As Aston stated in his paper:

> The General Post Office, the Custom House, and the Four Courts lie in ruins. They must presumably be rebuilt upon their old sites or their functions transferred to new buildings elsewhere. A national parliament house and group of modern Government buildings is urgently required. Is the already over-crowded College Green banking and commercial centre to be invaded, or is universal modern practice to be followed, and our new legislative assembly to be permitted to deliberate in suburban surroundings which are detached from the atmosphere and distractions of commercial activity? Is the General Post Office to be reconstructed upon a site which will stereotype the waste and inconvenience of carting postal matter backwards and forwards from four or five railway termini, or is the opportunity to be seized to combine postal and parcel traffic with centralised railway and steamer facilities? Are the Four

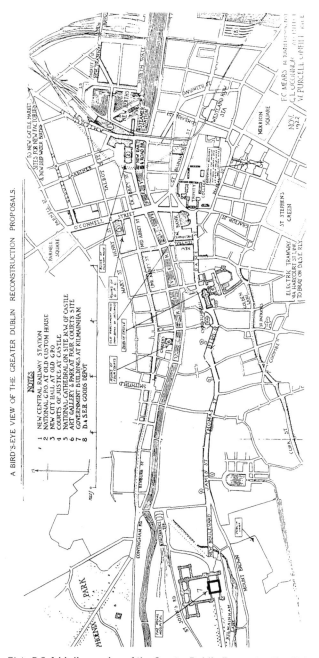

Plate 5.2 **A bird's eye view of the Greater Dublin Reconstruction Scheme, 1922**
City Archives, GRC/3/1/D

Table 5.1 **Key headings in the Greater Dublin Reconstruction Plan, 1922**

1	Reconstruction and the town plan
2	Existing administrative chaos
3	The housing question
4	Financial and other changes
5	Labour and 'reconstruction'
6	National buildings
7	Questions for corporation
8	Port problems
9	Transport
10	Suburban and rural problems
11	One problem! One authority?
12	History and topography
13	Eastward movement of city
14	Position of GPO and Central Station
15	Traffic centre of city
16	Proposed national highway
17	New city hall
18	New cathedral and courts of justice
19	Parliament House and government buildings
20	New north-western highway and municipal art gallery
21	Railways and southern quays
22	Tramways to north wall
23	New cattle market, foreshore reclamation
24	Other street improvements
25	Clearance of heavy goods
26	Parcels and luggage traffic
27	Electricity supply
28	Regeneration of western Dublin
29	Dublin and the greater Ireland
30	Scheme is severely practical
31	Design and execution of scheme – a suggestion
32	Acknowledgement of authors of scheme

Courts to be rebuilt as courts of justice, or is a more suitable site available? Is the present opportunity to be availed of to reach a decision upon a question which has agitated the public mind for many years, viz.; the provision of a site for a National Basilica, which is the ambition and intention of Catholic Ireland? Do we realise that the position of the new Cathedral must play a determining part in any plan of New Dublin, if civic and architectural considerations are to receive anything like their due weight?[27]

A central plank of the GDRM document was for a national highway that would stretch from O'Connell Street, through Westmoreland Street, Dame Street, and Thomas Street, 'to a new Parliament House and Government buildings on the site of the present Royal Hospital at Kilmainham in west Dublin' (plates 5.3 and 5.4).[28] This highway would lead to 'Ireland's new *Capitol*, placed in spacious and beautiful surroundings on the ancient site from which the Irish metropolis began its growth more than a thousand years ago.'[29] Along the route of the highway the GDRM envisaged the location of a number of important public buildings, from the new GPO on the site of the former Custom House, to the new City Hall and Municipal buildings on O'Connell Street (plate 5.5). The final stop along the route of the national highway was reserved for the most important building of all, the new Parliament House. Aston wrote of:

Passing through a new and spacious boulevard . . . [with] lovely woodland surroundings, in the middle of which stand the National Parliament House and Government buildings . . . A central and undisturbed feature of the capitol is the noble building designed by Sir Christopher Wren as a home for aged and disabled soldiers in the eighteenth century. Round it modern buildings have begun to arise, and will continue to extend as Ireland requires to increase her accommodation for the purposes of her central Government . . . On leaving the capitol complex we pass through Islandbridge into the Phoenix park, which we may either explore or leave for another day, finding our way to O'Connell Bridge by a new north-western thoroughfare passing at the back of the new Municipal Park and Art Gallery, which now faces the Liffey on the site of the old Four Courts and which retains the central hall and dome of that historic pile.[30]

The reconstruction movement also planned for the provision of a new National Basilica, 'flanked on one side by the spires of Christchurch Cathedral and on the other by the towers of the new Courts of Justice, which now occupy

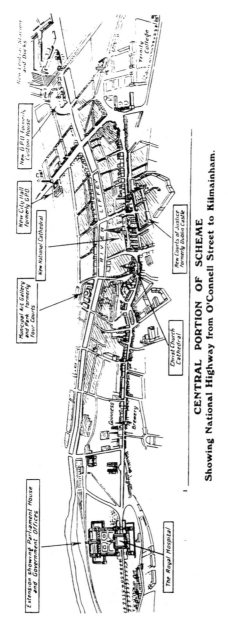

Plate 5.3 **The Greater Dublin Reconstruction Scheme, 1922**
Panoramic sketch looking northwards from St Patrick's Cathedral and illustrating the proposed
national highway.
City Archives, GRC/3/1/D

Plate 5.4 **Proposed new government buildings at Kilmainham**
City Archives, GRC/3/1/E

Plate 5.5 **Proposed central station and general post office**
City Archives, GRC/3/1/E

the site of what was formerly Dublin Castle. A striking symbol of the association of human justice with the religious aspirations of Catholic and Protestant Ireland!'[31] (plate 5.6) Despite lengthy discussion, few of the proposals of the reconstruction movement were actually implemented. But by the discussion of such large-scale changes in Dublin's built environment they rekindled in the public consciousness a recognition that development proposals should conform to a plan which was derived from a thorough Civic Survey. They also engendered much discussion, particularly on the issue of the Roman Catholic Cathedral and National Parliament building.[32] In the meantime, the proposal to house the government was the subject of much debate in the Dáil's temporary accommodation on Kildare Street.

Debating the power of place in Dáil Éireann

The GDRM proposal that the government buildings be housed out of the city in the west Dublin site of the Royal Hospital at Kilmainham attracted much attention in contemporary newspapers and journals. The *Irish Builder* championed an alternative proposal that the old Parliament House on College Green would be the most appropriate location. The return of government to the old site was seen by many as a symbolic reclamation of the space associated with previous legislative independence. Dublin Castle, which had escaped destruction during the years of civic unrest, and which had been for many years the bastion of British rule in Ireland was also considered as a possible location.[33] In July 1923 the first of two joint committees of the Dáil and Seanad was appointed to consider the issue.[34] In the contentious debates that followed TDs such as Alfie Byrne were forthright in their views:

> A considerable canvas is being made and a press propaganda has been arranged by certain individuals for a transfer of the Headquarters of the Oireachtas out to Kilmainham . . . Now I hold that the centre of the city is the proper site – no matter what property is involved in Kilmainham, and notwithstanding canvassing by people having property out there or people who do not want their property touched in the vicinity of what is now known as the Bank of Ireland . . . I hope the country if it is made clear at an election will get an opportunity of saying whether the Bank of Ireland should be the Headquarters of the Oireachtas or not.[35]

In his response to Byrne, President Cosgrave observed that he would not 'stand for huge expenditure for a Parliament House, if the Parliament can be

Plate 5.6 **The National Basilica**
View from O'Connell Bridge
City Archives, GRC/3/1/E

accommodated at a smaller price'.[36] Meanwhile, a report in the *Irish Builder* drew attention to the fact that 'The Royal Hospital would for many reasons be most unsuitable for a Parliament House', and continued 'We do not know on what grounds the President has been advised that the old Parliament House would be inconvenient. It seems to us on the contrary it would be eminently suitable, central, dignified, and with imperishable traditions.'[37] Similar sentiments were echoed in the *Evening Telegraph*:

> Isn't history a disappointing thing when we start making it in this country. There's the old House on College Green for which we fought for I don't know whether it is seven centuries or seven generations and as soon as we get it we won't have it. I'm sorry. When I recall all the times the flag was planted on the old House from every platform in Ireland, Europe, Asia, Africa, America and Cork, and I think now that Our Flag will ne'er be seen floating high o'er College Green I cannot but grieve. History is certainly ungrateful and after all we've done for it too. It's a bad day for Ireland when we put our sentiment into a bank. That's what we've done by declining to plant the flag on the old house.[38]

Nevertheless, senior government members favoured the site at the Royal Hospital, Kilmainham, on the grounds that they were seeking a suitable site, at short notice, and with the least possible expense.[39] A commission was then appointed by the Dáil to investigate the options for a permanent location but was unable to agree on any recommendation regarding temporary accommodation.[40] The commission report was eventually presented on 8 August 1923, although consideration was suspended until after the impending general elections.[41] By the end of 1923 a columnist in the *Irish Builder* remarked:

> We have excellent reasons for believing that the proposal to transfer the meeting place of the Irish Free State Parliament to Kilmainham has been definitely abandoned. The feeling of the country has been wholly against this proposition. The probabilities are that the Senate and the Dáil will continue to meet in their present temporary quarters until such time as a permanent home worthy of the Irish capital is made available. No other place presents such attractions as the old Parliament House in College Green, but of course, there are difficulties, mainly financial, in the way of providing the Bank of Ireland with another home.[42]

Despite the 'excellent reasons' held by the columnist in the *Irish Builder*, hopes were dashed when the president outlined the plans of government in January 1924:

1 That immediate steps be taken to have the buildings of the Royal Hospital, Kilmainham, made available for the temporary accommodation of the Oireachtas and the buildings at present occupied restored to their original purposes.

2 That a Committee of members of the Oireachtas be appointed to enquire and report to both Houses of the Oireachtas as to suitable and available sites for the permanent housing of the Oireachtas.

3 That the thanks of Dáil Éireann be and are hereby voted to the Royal Dublin Society for the facilities which have been placed at the disposal of the Oireachtas by the Society.

4 That a message be sent to Seanad Éireann requesting that this resolution be considered by them at their earliest convenience.[43]

A lengthy debate ensued with the president arguing that the Royal Hospital at Kilmainham was the best option open to the Oireachtas, if only on financial grounds. He put forward his argument 'in good faith, on the understanding that it is temporary accommodation and that we are not now in a position to spend huge sums of money upon housing even the first institution of the country in surroundings that would be fit for such housing'.[44] An amendment put by Darrell Figgis TD highlighted an alternative position. He stressed that, 'there is not only one, but more than one effective alternative . . . I only wish to recognise that Kilmainham is not the only alternative . . . it is still worth considering whether we should or should not continue here, and whether the Royal Dublin Society could not be given provision other than would be given to it by our relinquishing these premises.'[45] Figgis went on to argue strongly for a committee to be set up to consider the issue in greater detail and to recognise the strong sentimental feeling that existed for use of the old Parliament House as well as the antipathy towards the use of the Royal Hospital site. He closed by observing that, 'I think it is very desirable that the Parliament of the Free State should be somewhere right in the very centre of the capital of the State.'[46]

The debate on the temporary accommodation came before the Dáil once again in March 1924 when the report of the joint committee was issued.[47] It concluded that the only feasible buildings for the temporary accommodation of the Oireachtas were Leinster House, the Royal Hospital, Kilmainham, or Dublin Castle. While the Royal Hospital and Dublin Castle would afford ample housing accommodation, it could not recommend the selection of the Royal Hospital 'owing to the existing difficulties of access by the business and professional life of the City'.[48] The committee went on to note the prohibitive costs of the move and the fact that such a move could not take place for up to a year during which time Leinster House would continue to be occupied. The Castle was rejected on the grounds that although excellently situated it was already the location of the Courts of Justice. While the committee opted for the Leinster House site for the temporary Oireachtas accommodation, consideration of the report was suspended until 26 March 1924, after Deputy Thrift called for an examination of all the possible locations for the Parliament. He pointed to the possibility of housing the parliament in temporary buildings on Leinster Lawn while Deputy McBride suggested that:

> The capital of a country ought to be, as the capital of every other country of the world is, in the centre of the country. Dublin is really a foreign town. The streets, as you pass along, speak of the foreigner and of the foreigner's power. The capital of this country should be removed from the atmosphere of Dublin City. The Dublin people are courteous and very nice, but the seat of the government of this country should be far removed from the atmosphere of Dublin and from the centre of foreign power and from its foreign mode and

method of thought. We have heard a lot in this assembly of the reincarnation of the Gael and of the Gaelic State. The front bench talked about the reincarnation of the Gaelic State; we are going to start from the beginning . . . I think the seat of the government of this country should not be in Dublin at all.[49]

Thrift's amendment was also supported by Sir James Craig who argued in favour of the amendment for different reasons:

I feel we are not very comfortable here. Every night when I go home I have to put cotton wool around my shins in the same way that you would bandage a horse in preparation for a hurdle race. I do not think that this theatre is a fit place for an assembly of this kind. As I understand it, the suggestion of going to Leinster Lawn is a practical proposition. We could have plenty of room there.[50]

Although the motion was carried and the report was referred back for further consideration, this only prolonged the resolution of the issue.[51] The second report of the joint committee was issued in June 1924 and it confirmed that the parliament house would in fact remain at Leinster House.[52]

Between 1924 and 1926 Leinster House was converted for use as the Oireachtas, while the RDS gave up the house in return for a payment of £68,000 and was relocated to Ballsbridge. So, although there existed a range of plans for a new parliamentary complex, some more imaginative and overtly symbolic than others, the economic difficulties and practical realities of life in post-Independence Ireland made this an option too costly to implement. Unlike some other postcolonial capitals the new government did not build a symbol of independence but rather refurbished an eighteenth-century town house to fulfil the immediate needs of the new administration. Nonetheless, the entire debate is revealing of the potential role of the parliamentary complex in articulating national identity. The idea that Ireland's capitol should have been located at Kilmainham, 'from where the Irish metropolis began its growth more than a thousand years ago', or that it should have reclaimed the site of the former Parliament House in College Green highlights the symbolic role of architecture in contributing to the nation-building process, a contribution made even more potent when conjoined with being a site of parliamentary democracy. This was reiterated in Patrick Abercrombie's plan for the city that was published in 1922.

'Dublin of the future': Abercrombie's plan for Dublin, 1922

To accomplish the upbuilding of Dublin as an exemplary Metropolis a city plan for 'Greater Dublin' is a necessity. . . it is the desire of the members of the Institute to arouse the historic and traditional spirit of Civic pride once so evident, and to revive that native genius which will place Dublin in its proper position as one of the world's best Capital Cities, famed for the wholesomeness of its laws, the comeliness of its inhabitants, and the dignity of its labours.[53]

The re-establishment of a national parliament should give the necessary impetus to set a great town plan in motion, and the access of material prosperity which will ensue will provide the means to carry it out.[54]

While the proposals of the GDRM were issued in the aftermath of the 1916 Rising and subsequent wars, another planning development which pre-dates this, although not published until 1922, was Abercrombie, Kelly and Kelly's *Dublin of the future*. This plan was the winning design in the Dublin Town Planning Competition which had been launched in 1914 by the Lord Lieutenant, Lord Aberdeen.[55] The brief specified that the planning proposals were to relate to the greater Dublin area, taking in Howth, Glasnevin, Ashtown, Dundrum, and Dalkey and should have regard to three main headings, as illustrated in table 5.2 below. The competition attracted eight entries, four from Irish entrants, two from Liverpool and one from both London and Illinois, USA, which were assessed by a team of three assessors, Patrick Geddes, John Nolen and the Dublin city architect, Charles McCarthy. The winning entry was announced in August of 1916. The judges decided to award the prize of £500 to Professor Patrick Abercrombie, who had completed the winning entry with his colleagues Sydney and Arthur Kelly.

Although the winning entry had been announced in 1916 it was not until 1922 that the plan was actually published, by which time the city had endured the traumatic period of uprising and destruction that rendered *Dublin of the future* somewhat out of date. As Abercrombie et al. acknowledge in their introduction, it was:

a bold course to publish to-day a scheme for City Planning, for Dublin, which was originally prepared in 1914 – both in general and in particular. In general because, to those not yet well versed in the practice of Town Planning, a matter of eight years might well be thought enough to render the scheme somewhat antiquated. Though the particular scheme may be obsolete in point of detail, few towns have not suffered a change, physical and psychological, during these

Table 5.2 **Specifications for the 1914 Dublin town plan**

I   Communications

Road, rail and canal systems; existing and proposed industrial locations and main existing and proposed streets and thoroughfares

II   Housing

Existing tenements, number, locations and types of dwellings required, housing densities, phasing of development, open space using the American standard of ten per cent open space, and the relocation of institutions from the centre city to suburbs

III   Metropolitan Improvements

Better use of city's situation including the rivers and bay, preservation or expansion needs of public buildings (e.g. administration buildings, new art gallery, cathedral), provision of parkways and park system.

Compiled from P. Abercrombie et al., *Dublin of the future* (Dublin, 1922).

intervening years of war, trade boom and subsequent depression; but Dublin has added the double tragedy of war and civil war within her gates.[56]

The plan nevertheless provides a number of insights not only into how the greater Dublin area was to be planned but also as to how contemporary planners sought to incorporate a number of striking metropolitan improvements into the new capital of the independent state. It is important to mention, however, that, as the authors stress, the competitive circumstances in which the plan was conceived ensured that it was inevitably 'prone to produce certain faults which a maturer consideration would wish to remove. The spectacular is impossible to avoid when competing against others: it is necessary to arrest the eye with features whose boldness are perhaps more evident that their practicality.'[57] This point is borne out in some of the proposals which were put forward for metropolitan improvements (figure 5.1).

The plan as published was a detailed document covering a comprehensive range of issues. Primarily it provided for the complete reworking of the capital's inner city, ensuring that the city centre would in effect become a traffic hub (plate 5.7). An integrated transport system, focused upon the new traffic centre, was to be located on both sides of the River Liffey, immediately east of the Four Courts, and a new central station was planned for the riverside.

The creation of the traffic centre moved the focus of the city westwards, and left Sackville (O'Connell) Street as a monumental route-way, which would cut across O'Connell Bridge and down Westmoreland Street to the new Houses of Parliament. Further east, the Custom House was to become a new riverside station (plate 5.8). The overhead connection between Westland Row and Amiens Street, which disfigured the view of the Custom House, was to be replaced by an underground line.

Another element of *Dublin of the future* was the creation of the Phoenix Park Mall 'perhaps one of the most notable improvements in Dublin'.[58] The authors argued that:

> The possession of the finest town park in Europe should certainly stimulate a city to provide it with a worthy approach. It will be recalled that Gandon suggested that the Wellington monument should take the form of a triumphal arch at the entrance to the Phoenix Park. Unfortunately his sound advice was not followed, and an unrelated obelisk was set up in the park instead. It has been one of the chief concerns of the present proposal to bring this obelisk into due relation with the city, and fortunately it has been found possible to orientate the wide approach from the central square on its axis, the route from the Custom House . . . to Phoenix Park would be one of great grandeur.[59]

Dublin of the future also included plans for a national theatre to be built at the head of Sackville (O'Connell) Street, while the Rotunda Hospital, balanced by a similar structure on the other side, would be given over to new uses (plate 5.9). Abercrombie suggested that 'the Maternity Hospital would be given a rest in the country air with what was best in it to be preserved and the new wing on Granby Row rebuilt as a Music Auditorium'.[60] The whole group, which was to be combined with a city restaurant, was designed to face on to the Rotunda Gardens. The Royal Barracks was to become the site for a new national art gallery, while the newly formed site between the two avenues leading to Christchurch would form an appropriate location for a bourse, close to the banking headquarters of Dame Street.

Dublin of the future also incorporated plans for a new cathedral, which would face down Capel Street near the city centre. The basilica was designed to follow the style of the early Christian architecture of Rome, but incorporated 'a single lofty colossal shaft, founded on the traditional Irish Round Tower [which] would serve as the spiritual emblem of the city' (plates 5.10 and 5.11).[61]

The issue of building a new cathedral proved to be a contentious one which attracted much attention. It seemed to many in independent Ireland as

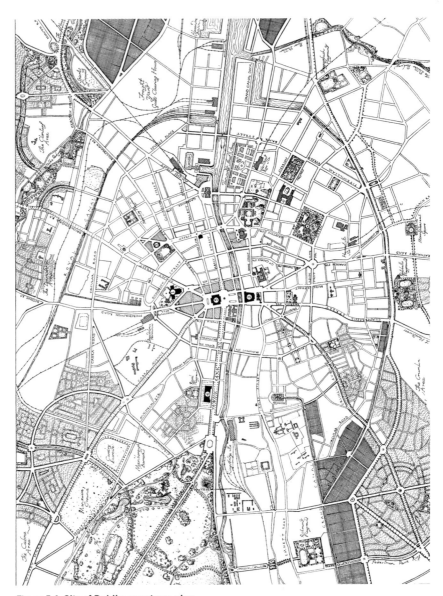

Figure 5.1 **City of Dublin, new town plan**

Reference key to suggested building locations:

1	Cathedral	4	Art Gallery
2	Parliament Building	5	Central Station
3	National Theatre	6	Riverside Station

From P. Abercrombie et al., *Dublin of the future* (Dublin, 1922), plate 5

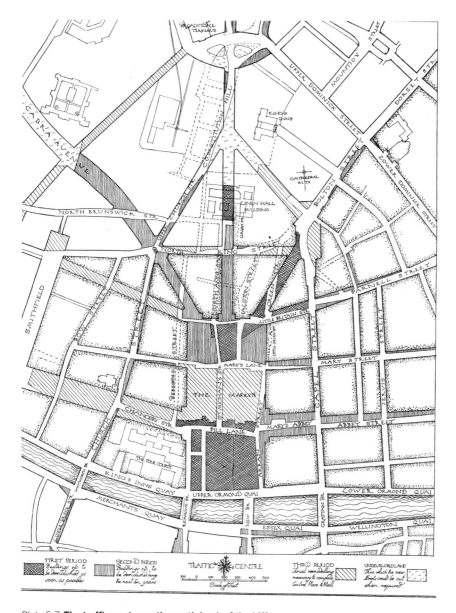

Plate 5.7 **The traffic centre on the north bank of the Liffey**

From P. Abercrombie et al., *Dublin of the future* (Dublin, 1922), plate 27.

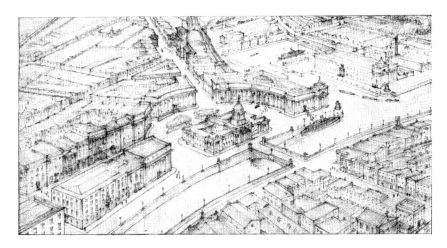

Plate 5.8 **Custom House crescent and riverside station**
The sketch shows the railway across the Liffey taken underground; the Butt Bridge removed;
a new road bridge facing the Custom House; the Riverside station on the site of the old dock and
the Docks remodelled for the passenger steamers; the crescent completed; and the approach to
Amiens Street improved.
From P. Abercrombie et al., *Dublin of the future* (Dublin, 1922), plate 11.

something of an anomaly that, given the religious composition of the population, the capital did not possess a Roman Catholic cathedral. Although a Pro-Cathedral had been erected at the beginning of the nineteenth century, it was built in an inauspicious location on the comparatively quiet thoroughfare of Marlborough Street. In fact, the issue of building a cathedral stretched back to 1914, when the Dublin town planning competition prompted the proposal that a cathedral be situated next to the Four Courts on Ormond Square.[62] The destruction of the city's chief public buildings during the period from 1916 through to 1922, however, opened up the options in terms of a cathedral site. A suggestion subsequently arose that it be erected on the site of the Four Courts, in a gesture which would have replaced what had become a symbol of British justice with what was popularly considered to be a more appropriate edifice for the new state.[63] An alternative site on Sackville Street also garnered much support, most especially in the pages of the *Irish Builder*. It lamented the lack of a cathedral, arguing that, 'the building in Marlborough Street is only a Pro-cathedral, not at all worthy of being considered in any sense a national basilica'.[64] It went on to suggest that:

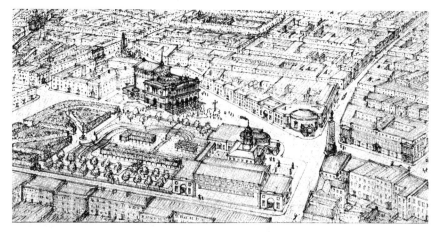

The National Theatre at the top of O'Connell Street. In the foreground is the new Auditorium and restored Rotunda ; the centre block is to be a restaurant facing on to the Rotunda Gardens or Rutland Square.

Plate 5.9 **The national theatre and auditorium**
From P. Abercrombie et al., *Dublin of the future* (Dublin, 1922), plate 33

The newly destroyed area opens up the possibility of a splendid site in the very best possible position. The frontage of the Pro-cathedral and the Presbytery in Marlborough Street could easily be enlarged at no great cost: this could then be opened out to Sackville Street, forming a fine quadrangle. Such a quadrangle could be disposed more or less axially in agreement with the finely disposed grounds of Tyrone House, and would be capable of being magnificently laid out. This site would afford an unrivalled site for the erection of a great and splendid Cathedral, emblematical of the hopes and aspirations of Ireland in the new era and architecturally affording some compensation for the grievous losses sustained by the destruction of the other great buildings, which were part of the pride and glory of Dublin.[65]

The cathedral issue was resurrected towards the end of the year in the *Irish Catholic*. The reporter observed the 'splendid opportunity' now afforded in Dublin 'to pick up the chain of an interrupted tradition by the erection of a really spacious cathedral which would represent or embody what has already been realised in the national style, while carrying it further by the introduction of domes or cupolas. Such an edifice would present great possibilities for the exercise of Irish decorative art'.[66] Despite the media and general interest in the issue, however, the proposal, like that for the new parliament building, did not

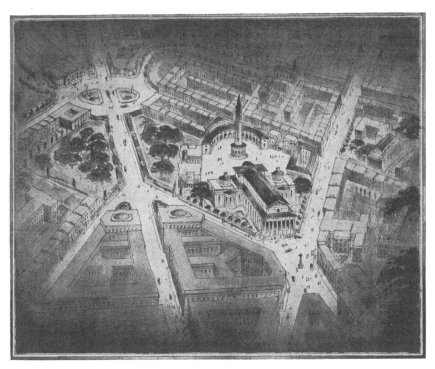

Plate 5.10

Diagrammatic sketch showing the suggested Cathedral site

From P. Abercrombie et al., *Dublin of the future* (Dublin, 1922), plate 32

Plate 5.11 **The proposed campanile**
From P. Abercrombie et al., *Dublin of the future*
(Dublin, 1922), plate 31

reach fruition but instead remained on the drawing board. In 1930, the Catholic Church acquired Merrion Square as a site for a new cathedral. The then Archbishop, Dr Byrne, announced on 12 July of that year that the park had been purchased from the Earl of Pembroke and Montgomery for £10,000.[67] The proposition met with a broad welcome. The *Irish Builder* for example observed:

> At last a noble site has been acquired which makes the erection of a cathedral no longer an impossible dream. Merrion Square is, indeed, a noble site for the purpose. Few cathedrals in Europe can boast such a one, and it has been obtained at the very moderate price of £10,000 for 12 acres in the heart of the city.[68]

As Dublin expanded and it was found necessary to build churches in the suburbs to cater for the increasing population, the Cathedral proposal was eventually abandoned. The site was given over to the public in 1974 by the then Archbishop, Dr Dermot Ryan.[69]

Rebuilding the destroyed public buildings

While much consideration was given to the building of a national cathedral and to the housing of the national parliament, the fact remained that many of the city's chief public buildings lay in ruins and awaited either reconstruction or demolition. The *Irish Builder* reported:

> Unless we are seriously degenerate, we must have regard to the value of our old buildings. They are the priceless heritage of the past. Nothing we have pro-

duced for a hundred years can compare with them. Surely the government will disprove the charge that they are indifferent to our 18th century buildings! A sympathetic touch, an affectionate gesture towards old traditions in Ireland, would at once redeem us from the charge of indifference. This country is not like England, with its countless memorials of the past. A few here or there would hardly be missed. Our old memorials are all too few. Every one of them ought and must be conserved.[70]

These sentiments were echoed in several eloquently worded protests in contemporary journals against the demolition of Gandon's Dublin masterpieces.[71] Ultimately, all of the major public buildings – the General Post Office, the Custom House and the Four Courts – were in fact gradually rebuilt over the course of the ensuing decade.[72] In the Dáil in 1922 the President declared his intention to ensure that all portions of destroyed buildings would be rebuilt and noted that all necessary measures of preservation had been taken. It is significant that the reconstruction of these buildings, which had come to be seen in certain circles as symbols of British rule, formed an important part of government policy and considerable sums of what was scarce public money was expended in the process.

As a result of the 1916 Rising the General Post Office had been reduced to rubble and its function transferred in part to Dublin Castle.[73] It fell to the principal architect of the OPW, T. J. Byrne, with the assistance of others, to prepare the plans for its reconstruction.[74] His plans were sanctioned by the Free State government in 1924, a decision which was greeted with much acclaim and was:

> welcomed by the vast majority of the public, as the most natural, proper, economical and common sense decision that could possibly have been taken. Here we have a building dedicated these one hundred and ten years past to the purpose of a General Post Office in the very heart and centre of the city, an admirable architectural design and one of the great relics of our eighteenth century civilisation happily left in such a state as to be capable of economical restoration, and at the same time inwardly so completely cleared out as to lend itself to the most modern scheme of replanning that could be desired.[75]

The building was finished in 1933 whereupon it was reopened to the public.[76]

The rebuilding of the capital's Custom House began in 1925 following a debate which raged regarding its future function.[77] The programme of rebuilding the Custom House by a combination of contract work and direct labour began in 1925 when contracts were let for the eastern, southern and western blocks.[78] During 1925–6:

> good progress was made on three contracts for reinstating to modified plans the east, south, and west blocks, these being roofed. As in the Four Courts the new floors have been put in concrete on structural steel increasing the light and air space, improving the office accommodation and simplifying the planning. Work on the north block has proceeded by direct labour, while much work in the relief of unemployment was given by continuous clearance in the centre blocks of the site.[79]

This work, which involved substantial remodelling of the interior, was largely completed by 1927 when the building was partly occupied.[80] The restoration was completed in the 1928–9 financial year at a total cost of £300,000. It is notable that in rebuilding the dome, native Ardbraccan limestone was used, significantly different from the British Portland stone that had characterised the earlier dome.[81]

In June 1922 the Four Courts which was severely damaged by fire and explosion to a much greater extent than had been the case during its occupation in 1916. The Four Courts and northern block were completely gutted by fire while more than half of the Record Treasury was destroyed and the Land Registry was also damaged. In fact, 'the damage to the buildings at this period was so severe that in the opinion of many experts demolition of the ruins was an essential preliminary to reconstruction'.[82] As a temporary measure the judiciary was transferred to Dublin Castle, while many impassioned pleas were issued for the rebuilding of the destroyed building, others argued that the site would be ideal for the proposed new Catholic cathedral.[83]

The Builder magazine, published in London, made its views clear on the matter in July 1922:

> It is not our custom to comment upon political affairs but when direct action is taken to the extent of causing the destruction of one of the most beautiful buildings in the world we cannot remain silent. Dublin contains unique examples of the finest eighteenth century work, among those the Custom House, recently burnt down, and the Four Courts are pre-eminent. The destruction of the latter must rank among the worst outrages in the history of architecture. The responsibility lies with those who seized the building and held it as a hostage, shielding themselves behind its beauty . . . we can only hope that the

Four Courts is not beyond repair. One of the first acts of the new Ireland on restoration of order should be the rebuilding of the Custom House and Four Courts and we sincerely hope they will be re-erected as they were . . . it would be a calamity to the world if they are to be lost to us in their original form.[84]

The *Irish Builder* also advocated the rebuilding of the Four Courts, suggesting that it was 'the bounden duty of the Government as soon as peace has been restored to institute a searching inquiry as to the possibilities of the restoration of the three great buildings which have of late so grievously suffered'.[85] After a period of some doubt and uncertainty it was eventually resolved in 1925 to rebuild. Cognisance was also given to the inherent meaning attached to the site as a court of law and that while it may have been the site of 'British justice', so too was it a magnificent example of Irish craftsmanship of the highest standard.[86]

In the mid-1920s rebuilding commenced under the direction of Mr T. J. Byrne.[87] Initial clearing had begun in 1922 and only a shell remained when Byrne took over. Although all of Gandon's drawings remained in the King's Inns ensuring that it was possible to undertake a complete reconstruction, Byrne argued that the building as initially conceived no longer fulfilled contemporary requirements. Hence, a new design was drawn up and, although the façade was retained, the interior was substantially remodelled.[88] Granite and Portland stone salvaged from the damaged walls were used in the reconstruction. During the rebuilding process, in a gesture of some symbolism, the emblems of Justice and Law which surmounted the triumphal arches on either side of the central block of the building were replaced by balls, while in the centre an Irish harp was put in place bearing the legend of the Irish Free State. Also, substantial use was made of native materials with every 'opportunity . . . taken to use local granite and sandstone from Kerry for all the new stonework and all the slated roofs are covered with slates obtained from Killaloe and Carrick-on-Suir'.[89] By 1930 all structural work was completed in the Supreme Court block and other contracts were being completed or sent out.[90] On 12 October 1931 the building was opened to the public in a ceremony which marked the first sitting of the Supreme Court and at which it was noted that:

We have been accustomed to hear sung the praises of the English architect, Gandon, who, during the time of his exile here, left his mark on the city in a group of notable architectural monuments, but, for the future, when speaking of three, at least, of these great buildings, the name of Byrne must be inevitably linked with that of Gandon.[91]

Forging national identity: new architectural initiatives

First of all we must cleanse our minds of imitation, and in disengaging ourselves of antique manners, bring out work to simplicity and truth, shaping each building to necessity, and look for beauty in chastity and proportion. That is the task before all Irish art and is especially the task before Irish architecture . . . During the past thirty years there have been many buildings erected in Ireland. With a few notable exceptions, not many of them have been such as need awaken our pride. I would be a wealthy man if I were to receive a golden coin for every column that carried no weight, for every arch that bore no burden, in all these buildings. How many men of high character among architects are there who would scorn to tell a lie, the memory of which would not last an hour, yet are satisfied to tell lies in buildings that will continue for a century.[92]

The expensive reconstruction of the chief public buildings destroyed in the capital during years of civic unrest together with the rebuilding of O'Connell Street, as well as the housing of the Irish Parliament, were just some of the architectural tasks which faced the new administration. Overall, those public buildings that were rebuilt followed a similar style to the original eighteenth-century construction, although the interiors were substantially remodelled to follow modern structural practices. As regards new architectural work, the image that the state sought to exude in the architectural fabric of its capital city did to some extent mark the dawn of a new era. This was to be a cultural paradigm which, according to the rhetoric at least, would distinguish Ireland from Britain, as suggested by one of the newly appointed government ministers, Darrell Figgis.

However, while the fledgling state set about the process of nation building and while 'the convergence of independence with the emergence of the Modern Movement offered the State an interesting choice in respect of its architectural expression . . . the debate about architecture was, however limited, erudite, sophisticated and informed.'[93] As O'Laoire suggests, the ruralist/cosmopolitan dialectic which prevailed in Free State Ireland and which perceived cities as bastions of oppression and privilege, had important implications for the architectural development of the capital. In the first ten years of the Free State, although substantial rebuilding work did take place, 'vacillation between historic certainty and projected ideals – between "essentialism" and "epochalism" – resulted in inaction'.[94] Beyond the capital one important development which distinguished this era was the building of the Ardnacrusha power plant, or the 'Shannon Scheme' as it was popularly known. As a number of commentators have suggested it embodied the spirit of the new nation, 'this huge complex on the River Shannon brought home the fact that Ireland could be a cement rich country and therefore self-sufficient in the basic building material, concrete'.[95] Although the Irish Free State did not espouse a national style of architecture, making it difficult to define the independent capital by its architecture, in the 1930s plans commenced for the construction of a building to hold the offices for the Department of Industry and Commerce. These plans offer an insight into the potentially powerful role of the architectural medium in the nation-building process.

The Department of Industry and Commerce building, Kildare Street

In 1935 a competition was launched by the OPW for a new scheme to centralise the various branches of the Department of Industry and Commerce. The building was to be located on the vacant site of the former Maples Hotel in Kildare Street and, 'with a view to seeking new inspiration in designs and relieving the existing pressure of work in this Office . . . the competition was limited to architects of Irish nationality resident in this country and also to architects of other nationalities resident here for the previous ten years'.[96] Thirty-six designs were submitted by thirty-three competitors and the winner was announced on 28 December 1935.[97] The competition afforded the private architect 'an opportunity to mark the progress that architectural education has made in this country during the past half century and had afforded a generation of students, trained in our school of architecture, an opportunity of dealing with a practical problem as a change from studio exercises'.[98] This limitation was criticised in the *Irish Builder*, where it was argued that 'either Irish architects are fit to compete against those of any other country in the world, or else there is something which foreign architects could teach them'.[99]

The commission was awarded to the County Cork architectural firm of Boyd-Barrett.[100] Contemporary journals expressed confidence in the fact that 'his conception will add greatly to the architectural amenities of a thoroughfare already containing the fine group of buildings which include the National Library, the National Museum, and Leinster House – the home of the Free State Parliament'.[101] Boyd-Barrett's design for the building drew to a large extent on contemporary European developments in architecture. The Scandinavians had pioneered the style, as exemplified in J. S. Siren's Finnish parliament building, with which the new Irish building forms an interesting contrast.[102] The project was finally passed by the OPW in December 1936.[103] Substantial use of native materials was made in the construction. Granite was quarried at Ballyedmonduff, County Dublin, the limestone for the strings, cornices and carved panels came from Ballinasloe, County Galway, while the stone for the jambs of the two tall windows came from Ballybrew Quarry, County Wicklow (plate 5.12). An important component, which reflects the meaning and symbolism inherent in the project, was the sculptural relief work on the building facing Kildare Street. Although Boyd-Barrett submitted preliminary designs for this sculptural work, the OPW, in consultation with the Minister for Industry and Commerce, Sean McEntee, suggested that several sculptors should be invited to submit designs.[104] The commission was eventually awarded to Gabriel Hayes. Her proposals consisted of two main keystones of Éire and St Brendan the navigator, two courtyard keystones, together with panels over the entrance door and under both the staircase window and the Minister's balcony. (plate 5.13).

The building was eventually completed in October 1942 and was occupied the following month. As the first commissioned public building of the Free State it stood as a symbol of the new era in Ireland. As Ryan puts it, 'At once classicist and modern, massive and modest, the Department of Industry and Commerce is an interestingly didactic representative of the fledgling new Ireland . . . in this era of the post-modern, the department . . . presents itself as a functionalist mass acknowledging its context and brief with a sober vigour.'[105]

The Shamrock Building at the World Trade Fair, 1939

Another architectural project which actually took place outside Dublin but which is important to consider in the context of post-independence nation building is the 'Shamrock Building', 'the most significant modern building never built in Ireland' – a vigorous modernist embodiment to the aspirations of the new state'.[106] A 1,200 acre site on a reclaimed marine marsh at Flushing Bay, Long Island in New York was the venue for the World Trade Fair which took place from April to December 1939. The trade fair gave participating countries the opportunity to 'show their wares' and its visitors the chance to tour the world in eighty minutes. Ireland had a place in this tour of the world where it was represented for the first time as an independent entity.[107] Each of the foreign pavilions, which were all heterogeneous in design and layout, were grouped around the US federal building and the lagoon of nations. Ireland's place amid the 'Lagoon of nations' was located some distance from the pavilion of the British Empire and took the form of a shamrock design ground plan designed by Michael Scott.[108] The government was concerned to ensure that

Plate 5.12 **Industry and Commerce building, Kildare Street**
Designed by J. R. Boyd Barrett, opened 1942

Plate 5.13 **Industry and Commerce building, Kildare Street**
Entrance door panel

the building would have a strong and recognisable Irish character particularly given that it was Ireland's first appearance as an independent state at an international trade fair, while it was also important to appeal to the twenty-five million Irish Americans.[109] Scott eventually settled on a design which, although drawing on a traditionally Irish motif – the shamrock – was very much in keeping with the theme of the exhibition, 'A new world of tomorrow'. As he later stated:

> I had a symbol of Ireland in plan and in elevation a modern building of our time. It would ring all the right bells. I thought of bee-hive huts, but you wouldn't know what they were from the air. And then the terrible thought came into my head: could I do a shamrock? At first I dismissed it. Then the more I thought of it, the more I thought it mightn't be bad in plan, so I made various sketches.[110]

Scott's building brought Ireland to an international stage. It was a fully fledged modernist design, heavily influenced by the work of Le Corbusier and with few concessions made to ornament. The other major features of his design included the use of the word 'Ireland' on the end face (plate 5.14). This had been a subject of some contention. Scott had initially favoured the use of 'Éire' but this was altered just before he left for America when de Valera sent a message ordering that it must be Ireland and not Éire. Although the latter term had been adopted in de Valera's 1937 constitution it was unfamiliar to many Americans.[111] Beneath the lettering stood a symbolic sculpture by Herkener of a young woman rising with her hands over her head from the sea.

The pavilion was opened on 13 May 1939 by Seán T. O'Kelly, the Tánaiste and Minster for Local Government and Public Health, who officiated in place of the Taoiseach. He delivered a speech on the significance of the World Fair for Ireland and expressed his eagerness for Ireland to take her place in the world of tomorrow and to do her part towards making the world a better place to live in. The Mayor of New York, Fiorella H. La Guardia, then:

> sent greetings to Ireland, to its scholarly president and to Mr. de Valera, Ireland's George Washington, whose confidence he was proud to say he enjoyed in the dark days of 1919. He sent greetings to all the Irish people whose kindred had contributed to the development of his great city of New York . . . pointing to a map of the six counties coloured red on a map of Ireland, the mayor, amid thunderous applause, declared his desire to see a united Ireland won by a united people.[112]

Plate 5.14 **The Irish pavilion at the World Trade Fair, 1939**
IAA

The Dublin Sketch Development Plan, 1941: architectural implications

The city of Dublin presents exceptional opportunities to the town planner. It is a capital city, standing in a magnificent natural setting; it possesses many great buildings and is laid out, in its major features, on the noble scale characteristic of the late eighteenth century; it also has the advantage of not having as yet grown excessively large.[113]

In concluding this examination of the architectural and planning initiatives that characterised Dublin after the formation of the Free State, it is interesting to explore the implications of the 1939 Dublin Draft Development Plan by Patrick Abercrombie, Sydney Kelly and Manning Robertson. It was and published two years later in 1941 as the *Dublin sketch development plan* (figure 5.2).[114] Their plan formed a marked contrast to the more outlandish proposals that were put forward in 1914. Since then, the economic realities of life in the Free State had clearly dented earlier optimism and this was reflected in a much more concise plan that embraced nine key areas (table 5.3).

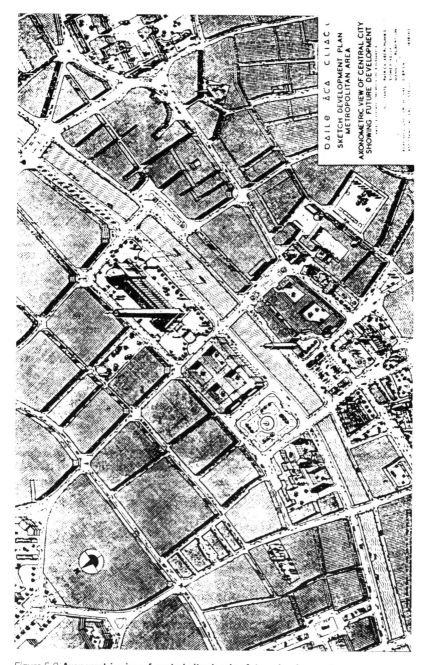

Figure 5.2 **Axonometric view of central city showing future development**
From P. Abercrombie, S. Kelly and M. Robertson, *Dublin sketch development plan* (Dublin, 1941), Map A, frontispiece

Table 5.3 **Key headings in the Dublin sketch development plan, 1941**

Communications
Regional planning
New public buildings
Housing and playgrounds in central areas
Neighbourhood centres
Health and recreation
Proposed reclamations
Zoning
Boundary of city

Compiled from P. Abercrombie et al, *Dublin sketch development plan* (Dublin, 1941)

Pertinent to the present discussion are the proposals which centred on civic improvement. The earlier preoccupation with superimposing a monumental traffic centre in the heart of the city had been replaced by a more subdued set of civic improvements, which set out nevertheless to make Dublin the focus of national pride and administration.[115]

> We have envisaged Dublin holding at its centre a great shrine and focus of public worship, and what could be more appropriate as a thanks offering to God for the restoration of peace when the blessings of peace shall reign once again in the world?[116]

The building of a new metropolitan cathedral was foremost among the proposals contained in the plan, to be designed as an edifice which would symbolise the spiritual life of the people.[117] The authors suggested that it occupy a central position which would lend itself well to ceremonial occasions as well as being easily accessible. '[I]n making a decision of such vital importance to the future of the city, it would be unwise to economise unduly upon site acquisition. A building of this nature, which will cost some millions of pounds to erect, is deserving of the expenditure necessary to ensure a proper site.'[118] Hence their suggestion that the building be carried out in stages and not on the site that had long been considered the most suitable, Merrion Square, but rather 'the rectangle bounded by Lower Ormond Quay, Lower Liffey Street, Capel Street and Upper Abbey Street'.[119]

If we visualise such conditions, we find here the ideal site for the greatest building and noblest monument of the future Dublin. Only one building presents itself which would fulfil these conditions, and that is the Metropolitan cathedral. No other building could fitly take its place as the climax to the City.[120]

A wide flight of steps was planned for the front of the cathedral which would face out onto the Liffey ensuring that:

The Cathedral would continue the Dublin tradition of facing its great buildings along the Quays. The Cathedral would occupy a central position, approximately half way between the Custom House and the Four Courts. Set in a green enclosure and bounded by widened streets it would be seen to the greatest advantage from across the river and from the neighbourhood of O'Connell Bridge.[121]

The dignity to be afforded the cathedral was provided for in the laying out of a proposed new thoroughfare from South Great George's Street, along with the opening up and widening of the Quays. A new bridge replacing the 'Metal Bridge' was designed to ensure that processions could approach the cathedral from either side.

In a subsequent article Abercrombie stressed that:

Many of the world's greatest buildings are set by the waterside. One can cite St Mark's and Santa Maria della Salute in Venice; Notre Dame in Paris, on its low-lying island; the great Town Hall at Stockholm; the Houses of Parliament in London and Budapest, and scores of others among the world's finest buildings. Few, if any, other cities are divided as geometrically by a fine river as is Dublin by the Liffey. All the main traffic routes will cross the river in view of the cathedral site, and the slight bends in the river will allow for the gradual unfolding of the building, from slightly different angles, as one proceeds down the quays.[122]

Moreover, he argued that, 'in the case of our proposed Cathedral site, there is no building on the whole site with any claim to architectural or historic merit or value . . . One cannot imagine that Irish people throughout the world would be prepared to subscribe to a cathedral building but not to the acquisition of a site upon which the whole Cathedral project hinges.'[123]

The building of the cathedral was but one architectural element of the plan. It was to take place in tandem with the regeneration of the city quays and would work in harmony with the proposals for a new Civic Centre on the south bank of the Liffey.[124] Designed to house administration for city and country, 'the civic centre should include the existing city hall. Parliament Street should be widened to 150 feet and the new Civic Buildings should flank the widened Parliament Street on the West, extending from Lord Edward Street to the Quay. The block should be l-shaped with a wing extending along Essex Quay. This wing would give dignity, sadly lacking at present, to the Quays on the south.'[125] The increased width of Parliament Street would enable it to be laid out with dual carriageways and be lined with grass borders. It was envisaged that the municipal buildings would form a centre for civic life atop the hill leading up to the existing City Hall.

The plan also incorporated a proposal for new government buildings which were in the 1940s housed in three separate places: Upper Merrion Street, the Custom House and the Castle. The authors suggested that the area around Merrion Square be 'colonised' for governmental use, stating that, 'the most suitable neighbourhood for additional buildings would, in our opinion, be that contained in the rectangle bounded by St. Stephen's Green North, Upper Merrion Street, Kildare Street and Leinster Street–Clare Street'.[126] They argued that it lent itself admirably to the requirements of a government centre which would be associated with the existing Oireachtas buildings, the National Library, National Gallery and Museums.

The Town Planning Committee of Dublin Corporation responded favourably to the proposals. They approved of the site selected by the consultants, stating that:

> In view of the inadequate and unsatisfactory accommodation in the existing scattered buildings used as Civic Offices, the Town Planning Committee agreed that new buildings are urgently necessary. It, therefore, recommends that steps be taken to invite designs for new municipal buildings by prize competition. The design should permit the erection of the buildings in sections. They should also provide for the inclusion of Council Chamber, Public Assembly Rooms, Lord Mayor's Office, etc., as well as the usual municipal offices.[127]

The *Irish Builder* echoed this positive response. The promise of a new river vista westwards from O'Connell Bridge, in which the new metropolitan cathedral was set to face an imposing pile of metropolitan municipal buildings, drew the admiration of one correspondent. It was argued that such buildings would form central and inspiring features and would:

surely make an irresistible appeal to citizens who are not entirely bereft of imagination. Likewise, and nearby, the demolition of structures which obstruct the view from the Liffey side of the Christchurch group of ancient and beautiful buildings should surely meet with ready popular acceptance . . . the suggestion that the city's west end . . . should become the permanent home of national administrative buildings in the vicinity of Leinster House, and with frontages to St. Stephen's Green, Merrion Street, Clare Street and Kildare Street, is a conception, in completest harmony with civic convenience and national dignity.[128]

In the journal *Studies*, three prominent individuals with an interest in planning issues were invited to comment on Abercrombie's plan. Louis Giron, the then president of the Royal Institute of the Architects of Ireland, observed with regard to the proposed Cathedral that 'the Merrion Square site was not entirely suitable'. It was 'somewhat remote from the centre of the city' and 'was not sufficiently dominating'. Owing 'to the nature of the approaches to the Square, a cathedral there could not be viewed as a whole to advantage', and it would 'be a pity to build on one of the city's open spaces, if an equally suitable site could be found elsewhere',[129] hence, his support for Abercombie's quayside site which would enable the construction of a cathedral visible for a considerable distance along the south bank of the river and in view from all of the most important city bridges.

In his response, the architect, John J. Robinson expressed a contrasting view. Although he did note that a cathedral 'worthy of our capital city and a symbol of the faith of our Fathers, placed upon the site in the surroundings visualised by the professor [would] satisfy even the most critical and would stand as a monument to the greatness of the concept which brought it about', he argued that the costs of levelling a quayside site would be exorbitant when placed in comparison with using the Merrion Square site.[130] Moreover, 'A fine cathedral, properly conceived and designed on noble lines, will dominate anywhere, even at the bottom of a deep valley, with buildings climbing on each side and all around it.'[131] He then proceeded to put forward an argument for the demolition of the Georgian houses on Merrion Square in an attempt to provide more space for the cathedral:

The Merrion Square houses, in spite of their gentlemanly and indeed genteel facades, are really very old gentlemen whose days are numbered. They are as moribund as their Capel Street contemporaries in spite of their more orderly old school tie traditions, and I venture to suggest that their term of survival is just as limited . . . their day is done – the Georgian era is over, and there is little sense in seeking to perpetuate it . . . their conversion into flats or offices makes only a poor compromise, however skilfully it may be carried out. The social structure and conditions which have brought them into being have long ceased to exist, and I submit that nothing is left for them but demolition – with possible modern replacement or the creation of open spaces.[132]

Robinson went even further by suggesting the demolition of the block of buildings between Upper and Lower Mount Street, Merrion Street and the Canal, which could be used for a spacious *piazza* with a great national monument in the centre 'and with the glorious mass of the new cathedral in the background'.[133] Even if the city's authorities were to clear the land from Merrion Street through to Fenian Street, Robinson argued that a new parliament house could be built on the site which would serve to add to the dignity and dominance of the nearby cathedral. He concluded with the question, 'why should all the grand civic planning be reserved for the north side of the river, when experience has shown that development tends southwards?'[134]

The final invited reaction to the Abercrombie plan came from C. P. Curran, who, in a detailed account, welcomed the plan for Dublin and argued for the preservation of all existing open space, including Merrion Square. He too favoured the riverine site for the new cathedral and also proposed the building of government offices adjacent to the Four Courts. He expressed regret at the fact that the plan assumed the permanent location of the Oireachtas buildings to be in the Leinster House area:

I had thought that this area of culture was a land permanently dedicated to the muses . . . it seems a pity that a sudden emergency should be the basis of a lasting development. I should certainly have preferred to see the new quarters of the Ministry of Industry and Commerce on a river site near the business area of Dame Street and near the Castle financiers. On the site in Kildare Street the Government would have earned our enduring thanks by the erection of the much needed concert hall.[135]

He suggested the riverside site would be 'opposite the new spectacular improvement between Christchurch and the river, it would also establish a fine architectural relation between the Four Courts and the cathedrals and would link the new Cathedral with its earliest ecclesiastical and civic history.'[136]

Conclusion

Although the *Sketch development plan* encompassed a range of novel proposals for the municipal centre, few of these were carried out, at least in the immediate term. The outbreak of 'the Emergency', together with the economic environment that prevailed, prevented the proposals from reaching fruition. The Irish Free State had, nevertheless, made some strides in channelling the expression of its independent status. The new state had successfully rebuilt many of the destroyed areas and had begun a substantial programme of building schools, hospitals and libraries, as well as the Shannon Scheme. The most significant buildings that were erected included the Industry and Commerce building and Dublin Airport in north County Dublin. All reflected in some measure the independence of the fledgling state and the role of architecture in cultivating national identity, as did the construction of the Irish Pavilion at the World's Fair of 1939. The outbreak of war in 1939, however, plunged Ireland into a gloomy period that left the architect Raymond McGrath to remark in 1941:

> We stood on Cruagh and looked down over the grey, silent desolation of the city. A thread of silver river broke through the monotonous ruin, punctuated here and there by the stumps of bridges. One solitary monument, the Testimonial in Phoenix Park, stood up like an enormous headstone. For the rest, spireless, domeless and treeless stretched that stricken waster – from Blackrock to Castleknock, from Rathfarnham to Raheny.[137]

The declaration of the Irish Republic in April 1949 marked one of the most dramatic developments in the immediate post-war period in Irish politics.[138] In November 1948 the Republic of Ireland Bill was introduced into the Dáil and the Act came into force on Easter Monday 1949, a symbolically appropriate date, which marked the thirty-third anniversary of the 1916 Rising. This was an optimistic period of economic growth, a spin-off of both the increased consumer spending at home in the aftermath of the austerity of the war years and also the rise in exports to Britain.[139] The infrastructural building programme, that had been initiated in the decades following the achievement of independence continued, marked by suburban housing initiatives and large-scale road development.[140] However, the initial optimism that prevailed in the post-war republic soon evaporated in a decade that was marked by successive economic crises.

Amid the economic gloom of the period, which was arrested to a great extent with the programme of economic rejuvenation initiated by T. K. Whitaker, a secretary in the Department of Finance, consolidation of nationhood continued apace.[141] After the war, the language of modernism was introduced into Irish architecture and found expression in a number of significant public buildings, prominent among them the central bus station, Busárus, and Liberty Hall. These buildings indicated the dawn of an era that was characterised by 'a palpable air of rebirth and renewal'.[142] As McGrath commented, the 'delicious smell of Portland cement, wet timber and midnight oil', predominated.[143] The key buildings that were subsequently erected perhaps inevitably came to be seen as symbols of significant industrial and social development, 'inevitably entwined with a heated polemic about the direction of social change and the cultural identity of our built environment'.[144]

CHAPTER SIX

Scripting national memory: the power of public monuments

New sculpture for a new capital

When the Free State came into existence, Dublin's monumental landscape embodied the contested heritage of previous generations. The power of Dublin Corporation, a strongly nationalist body in the late nineteenth century, to sanction the erection of monuments ensured that there was no shortage of nationalist heroes already in place on the streets of the new capital. Indeed, O'Connell Street was almost entirely lined with such figures who stood in an uneasy juxtaposition with an earlier erected emblem of empire dedicated to Lord Nelson. The Rising of 1916, coupled with the War of Independence and Civil War, provided a host of new heroes to stand upon pedestals throughout the city and country. In their geography and iconography these monuments carved out a visible landscape of memory as a testament to the new political situation (figure 6.1 and table 6.1).

Figures like Michael Collins, Arthur Griffith and later Kevin O'Higgins were commemorated in a cenotaph erected at the rear of Leinster House. Other leading figures of the Rising were also to find positions of public prominence in the capital, among them Countess Markievicz to whom a monument was dedicated in St Stephen's Green in 1932 and Sean Heuston, who was commemorated in the grounds of the People's Gardens, Phoenix Park in 1943. On the fortieth anniversary of the Rising in 1956, the Custom House memorial was unveiled, in memory of the members of the Irish Republican Brotherhood who had died in the War of Independence when the Custom House came under attack on 25 May 1921. In the same year, a site on the north side of the city at Arbour Hill was given over to the formal commemoration of the dead of the 1916 Rebellion.[1] The Golden Jubilee 'celebrations' in April 1966 involved a two-week, countrywide commemorative celebration to honour those who took part in the Rising and 'to emphasise its importance as a decisive event in our history'.[2] While special ceremonies took place at provincial centres all around the country, a number of significant statues were unveiled in the capital along with the formal opening of the Garden of Remembrance, and the presentation of the statue of Robert Emmet to the Irish state.[3]

Table 6.1 **Statues erected, destroyed and removed in Dublin, 1922–66**

Individual	Date	Location	Sculptor
Cenotaph	1923	Leinster Lawn	Atkinson and Power
King William III	1928 bombed	College Green	Gibbons
	1929 removed		
Countess Markievicz	1932	St Stephen's Green	Power
Cúchulainn	1935	General Post Office	Power
Tom Kettle	1937	St Stephen's Green	Power
King George II	1937 bombed	St Stephen's Green	
King George I	1937 sold	Relocated to Barber Institute	
Irish National War Memorial	1940	Islandbridge	Lutyens
Sean Heuston	1942	Phoenix Park	Murphy
Thomas Davis foundation stone	1945	St Stephen's Green	McGrath[1]
Queen Victoria	1948 removed	Leinster House	Hughes
La Piéta	1948	Marlborough Street	E. Luppi
Civil War Obelisk	1950	Leinster Lawn	R. McGrath
O'Donovan Rossa	1954	St Stephen's Green	S. Murphy
Three Fates	1956	St Stephen's Green	J. Wackerle
Dublin Brigade	1956	Custom House	Y. Renard-Goulet
1916 Memorial	1956	Arbour Hill	OPW
Gough	1957 bombed	Phoenix Park	
Carlisle	1958 bombed	Phoenix Park	
Eglinton	1958 bombed	St Stephen's Green	
Thomas Davis	1966	College Green	O. Kelly
Na Fianna	1966	St Stephen's Green	A. J. Breen
Remembrance Garden	1966	Parnell Square	D. Hanly
Nelson	1966 bombed	O'Connell Street	

Compiled by Y. Whelan using OPW files, the *Irish Builder* and *The Irish Times*, 1922–66.
1   This monument was never completed. Another monument was erected on College Green in 1966, designed by Edward Delaney.

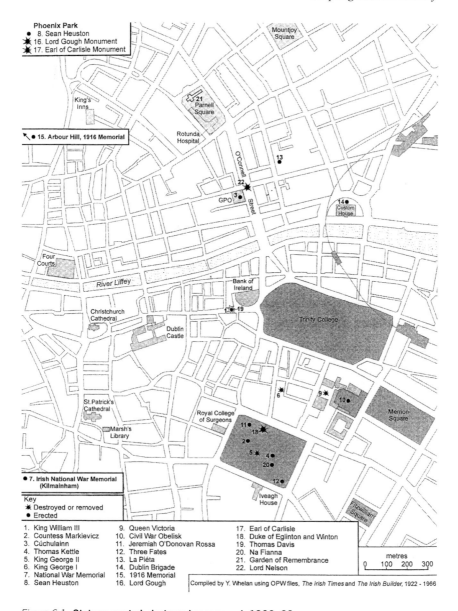

Phoenix Park
- 8. Sean Heuston
- ✹ 16. Lord Gough Monument
- ✹ 17. Earl of Carlisle Monument

King's
Inns

● 15. Arbour Hill, 1916 Memorial

Mountjoy
Square

21
Parnell
Square

Rotunda
Hospital

O'Connell Street

13

22✹
GPO 3●

14●
Custom
House

Four
Courts

River Liffey

Bank of
Ireland

Christchurch
Cathedral

Dublin
Castle

1✹ 19

Trinity College

6✹

9✹ 10●

St. Patrick's
Cathedral

Royal College
of Surgeons

Merrion
Square

Marsh's
Library

11● 18✹
2●
5✹ 4●
20●
12●

● 7. Irish National War Memorial
(Kilmainham)

Iveagh
House

Fitzwilliam Square

Key
✹ Destroyed or removed
● Erected

1. King William III	9. Queen Victoria	17. Earl of Carlisle
2. Countess Markievicz	10. Civil War Obelisk	18. Duke of Eglinton and Winton
3. Cúchulainn	11. Jeremiah O'Donovan Rossa	19. Thomas Davis
4. Thomas Kettle	12. Three Fates	20. Na Fianna
5. King George II	13. La Piéta	21. Garden of Remembrance
6. King George I	14. Dublin Brigade	22. Lord Nelson
7. National War Memorial	15. 1916 Memorial	
8. Sean Heuston	16. Lord Gough	

metres
0   100   200   300

Compiled by Y. Whelan using OPW files, *The Irish Times* and *The Irish Builder*, 1922 - 1966

Figure 6.1 **Statues erected, destroyed or removed, 1922–66**

The 1916 Rising was not, however, the only impetus behind public sculpture in Free State Dublin. On the contrary, other earlier rebellions, including those of 1798 and 1848, also shaped the monumental landscape. In September 1945, the centenary of the death of the Irish poet and patriot Thomas Davis was the catalyst for a display of Irish nationalism on the streets of the capital when the foundation stone of a statue dedicated to him was laid in St Stephen's Green. The Fenian leader and poet, Jeremiah O'Donovan Rossa was also commemorated with a monument on 6 June 1954. A monument of Robert Emmet was donated to the State in 1967 and erected on St Stephen's Green shortly after. Afforded prominent geographical positions, these monuments stood in contrast to the somewhat anachronistic monarchical figures erected some decades before and whose future was uncertain in the context of Free State Ireland.

Commemorating the Civil War

In the name of the Irish nation and by the Irish nation commissioned and empowered, we offer here a symbol of Ireland's reverence and sorrow, of Ireland's pride and gratitude, to the memory of two heroic men . . . A year ago, within the space of two weeks, came the awful tragedy of their death. They gave their lives in doing their duty to Ireland.[4]

It was with these words that the Irish president, W.T. Cosgrave, under a dull, leaden sky unveiled the memorial dedicated to Michael Collins and Arthur Griffith, two heroes of the struggle for independence and both victims of the Civil War. Erected at the rear of Leinster House, the seat of the Irish Parliament, in the centre of Leinster Lawn, the cenotaph marked the first formal commemorative memorial dedicated to those who had given their lives 'in doing duty to Ireland'. It was a monument which the new Free State government wasted little time in having erected. The commission was granted to Professor George Atkinson of the Dublin Metropolitan School of Art in 1923 and it was his brief to design a temporary memorial for a site at the rear of Leinster House in time for the first anniversary of the deaths of both men (plate 6.1).[5]

Atkinson designed a memorial in the form of a Celtic cross made of timber to a height of 12 metres, which in its artistic design linked the dead heroes with the Catholic and Gaelic tradition and fitted in with the 'Gaelicisation programme promulgated by the new administration'.[6] Two large pylons were erected on either side of the outsize cross, and plaster medallions of Griffith and Collins designed by the sculptor Albert Power were affixed to it. The words, 'Cum Glóire Dé agus Onóra na hÉireann', which translate as, 'For the glory of God, and the honour of Ireland', were inscribed in Gaelic script.[7] The location afforded the monument was also of some significance, given its position just in front of the statue of Prince Albert. The cenotaph can be seen to be a celebration of contemporary Irish heroes, as well as a proclamation of Ireland's independence. [8]

The progress of the monument and plans for its unveiling were closely followed in the contemporary newspapers.[9] The government put in place simple and brief arrangements for an unveiling ceremony at which the president was to deliver an oration from the base of the cenotaph, and issue a directive which called for flags to be flown at half-mast on all government buildings and military barracks. The plans also provided for an artillery salute of three volleys of four guns, each at intervals of one minute while the president unveiled the memorial.[10] The ceremony took place on 13 August 1923 when President Cosgrave made it clear that it was more than a tribute to the dead alone, rather the erection of the monument stood as proof:

> that the work to which they gave their lives has been carried to success, and that those who stepped into the gap of danger when they fell, remember with pride and reverence, the men from whom they derived their chief inspiration. Griffith and Collins died in a dark hour in the history of the State which they, more than anyone else, had helped to re-establish as a free nation . . . Griffith and Collins endowed this nation at once with liberty and with the power to use it.[11]

The day was marked by a huge gathering of the general public who assembled outside the railings and stretched out into thick lines from Leinster Street along Merrion Square as far as the entrance to Government Buildings:

> although they could only get a glimpse of proceedings they nevertheless looked on patiently and reverently. The whole spectacle was striking and imposing . . . The cenotaph stood in front of the great assemblage, but covered from top to bottom with a white cloth, held in position with ropes. The Last Post broke the silence for a moment as the President laid four massive memorial wreaths at the base of the cenotaph . . . great silence of four and a half minutes duration then fell upon the uncovered and upstanding assemblage and a spectacle of the most solemn and affecting kind was seen . . . eyes were moist, while lips quivered with prayers for the prized and lost ones.[12]

Plate 6.1 **The Cenotaph, Leinster Lawn**
Erected in 1923, designed by George Atkinson and Albert Power
From the Valentine Collection, NLI

From the time of its unveiling, the cenotaph acted as an important focal point for annual state ceremonies when wreaths were laid and military honours observed in honour of those who had lost their lives in the struggle to achieve independence. This practice was, however, discontinued from 1932 after de Valera came to power,[13] although a private wreath-laying ceremony did take place by the Army Comrades Association with the agreement of the government.[14] The election in 1932 of Eamon de Valera as Taoiseach heralded a somewhat ambivalent attitude towards public memorials dedicated to those who had died in the independence struggle, not least owing to the fact that in his eyes the struggle had not yet been completed. Despite the political differences between Fianna Fáil and Cumann na nGaedheal, it fell to the former administration to set about replacing what was always intended as a temporary cenotaph with a permanent memorial. Hence, in 1940, the government approved a design for a permanent cenotaph by H. J. Leask, Keeper of National Monuments. His initial drawings made use of the Celtic cross motif. With the outbreak of the Second

World War, however, the project stalled. It did not come to fruition until 1947 by which point Leask's plans had been abandoned in favour of an obelisk.

The issue of removing the temporary cenotaph and erecting a new memorial in a more central position on the lawn had always been envisaged, although in practice it proved something of a contentious issue.[15] Leask's proposal was subsequently dropped after the war and an alternative proposal was drawn up by the principal architect of the OPW, Raymond McGrath.[16] McGrath's design replaced Albert Power's Celtic cross with a 60-foot high granite obelisk made of Wicklow granite. It incorporated a circular sloping granite base with an inlaid cross of gold on the face. The monument was topped by a flaming torch, with three bronze medallions enclosed by laurel wreaths surrounding the base.[17] The monument was completed in 1950 but was never unveiled with an official ceremony, owing most likely to the divide that the Civil War had engendered in Irish politics. It nonetheless represented a powerful symbol of the new republic. What had begun as a relatively simple Celtic cross, an attempt by the new Cumann na nGaedheal administration to assert political legitimacy, was completed with an obelisk, a universal image of power, soaring into the sky (plate 6.2). It is of some significance that the monument was erected on the site formerly occupied by Prince Albert, which was moved to one side to make way for the obelisk.

Commemorating the 1916 Rising

As the cradle of the 1916 Rising, it might be expected that the capital's central thoroughfare, O'Connell Street would play a crucial role in the symbolic construction of national identity. Initially the authorities were preoccupied with the reconstruction of the large portions of the street that had been destroyed during the years of war. Once the process of reconstruction was set in train, O'Connell Street assumed a central role in the annual spectacle that marked the commemoration of the rebellion. Although the thoroughfare was already lined with an array of public statues, a monumental initiative inside the General Post Office did symbolise Ireland's emergence as an Independent State and commemorate all those who had been killed in the rising.

Plate 6.2 **The Obelisk, Leinster Lawn**
Erected in 1950, designed by Raymond McGrath, OPW

The 1916 memorial in the GPO

When Pearse summoned Cuchulain to his side,
What stalked through the Post Office? What intellect,
What calculation, number, measurement, replied?

From W. B. Yeats, 'The Statues', April 1938.[18]

One of the first public monuments erected in commemoration of those who had been killed in the 1916 Rising was unveiled on 21 April 1935 in the General Post Office. On that day, which marked the nineteenth anniversary of the Rising, thousands came out onto the streets of the capital to witness the military parade that accompanied the civil ceremonial and to hear President Eamon de Valera declare in his unveiling address 'that only an Ireland free from foreign domination – North, South, East and West – would satisfy the aspirations of the Irish people'.[19] The memorial was modelled by Oliver Sheppard in 1911–12 and took the form of a bronze sculpture of Cuchulainn, the heroic figure of Irish mythology who was championed by Irish nationalists as an exemplary Celtic hero (plate 6.3). The design was lauded by de Valera as an apt representation of the heroism of those who had been killed in the Rising. As Turpin points out, 'The legend of Cuchulainn, with its heroic ideals of service to one's people before one's self, and the evocation of an ancient and noble Irish society, appealed greatly to the romantic imaginations of Celtic revivalists . . . The saga could be seen as a challenge to Irish political subservience to England and to modern "materialistic" values.'[20] Moreover, the sculptor made use of the Pietà theme from religious art in order to represent the heroic figure in an almost Christ-like fashion, bravely meeting with death. The use of such imagery in a monument dedicated to those who had died in the rebellion was significant, juxtaposing as it did the ideals of Christianity with those of revolutionary nationalism to create a potent symbol that would contribute to the visible script of national identity. However, the chosen symbol was also an ambiguous one given that Cuchulainn was also championed as a loyal defender of Ulster by those whose ideology was at odds with that of the Free State's leaders. As a columnist in *The Irish Times* observed, it was 'somewhat paradoxical that the warrior who had held so long the gap of Ulster against the southern hordes should now be adopted as the symbol by those whose object it is to bend his native province to their will'.[21] This point was also alluded to in *The United Ireland Journal* where a columnist suggested that 'there is nothing told of Cuchulainn that would make a representation of his death a suitable symbol for the struggle and sacrifice of 1916'.[22]

Despite these reservations, the Government went ahead with plans to unveil the monument on the nineteenth anniversary of the Rising when it served as the centrepiece in the elaborate military display that marked the occasion. The roll of honour on the day included 1916 veterans – the President, de Valera, the Vice-President, Sean T. O'Kelly, and Sean Lemass – along with relatives of the seven signatories. Detachments from every section of the Free State Army and from every district in the country took part. The day began on a bleak note, however, with 'rain sodden tricolours that flapped dismally over Dublin's buildings . . . There was no great display of bunting and a queer empty feeling seemed to be present in us all as we watched the preparations made to commemorate the Easter Week Rising'.[23] While the crowds listened to a voice that recounted the exploits of Easter Week, company after company filed into O'Connell Street. Contemporary newspapers record that:

> On they came, marching in military columns of four, some of them wearing the uniforms of the period of the Rising which for years had been hidden against raiding parties of British soldiers. Others had a bandolier or a knap-sack –

anything that was 'a relic of 1916' – and on their shoulders the men who were to fire the salute bore the old Mauser rifles, landed during the Howth gun-running by the late Erskine Childers.[24]

They were followed by detachments of men who had fought in the GPO, Boland's Mills, Jacob's and other outposts of the rebellion and were accompanied by the women who had acted as nurses.

In his unveiling speech de Valera drew particular attention to the proclamation issued in 1916 and expressed the hope that it would 'serve to keep in the minds of the youth of this country the great deeds of those who went before us, and that it will also serve to spur us on to emulate their valour and their sacrifice'.[25] He also remarked upon the suitability of the site, noting that:

> From this place nineteen years ago the Republic of Ireland was proclaimed. This was the scene of an event which will ever be counted an epoch in our history – the beginning of one of Ireland's most glorious and sustained efforts for independence. It has been a reproach to us that the spot has remained so long unmarked. Today we remove the reproach. All who enter this hall henceforth will be reminded of the deed enacted here. A beautiful piece of sculpture, the creation of Irish genius, symbolising the dauntless courage and abiding constancy of our people, will commemorate it modestly, indeed, but fittingly.[26]

At the stroke of twelve o'clock the monument was finally unveiled and the military spectacle continued when the GPO was filled with the sounds of the trumpeters in the gallery and on the roof of the building. Veterans of the Rising also fired a salute which was answered my members of the Free State Army marshalled on O'Connell Bridge. The effect was such that 'the whole range of buildings in O'Connell Street resounded with the din. It recalled the noise that accompanied the deadly bullets which in 1916 flashed to and fro across this same street'.[27] While the Army Band struck up 'The Soldier's Song' the military march past began and de Valera left the GPO for a rostrum outside where he watched as over 6,000 members of the army, drawn from all parts of the country, marched by in an hour-long ceremony. The end of the display was signalled by three Air Corps aircraft who plunged downwards in 'V' formation before rising again.

The spectacle of unveiling did not pass without controversy, however, and the absence of members of the opposition benches as well as of more militant republicans threw into sharp relief the political tensions that prevailed in the

Free State, where 'Men who had been comrades in arms nineteen years ago now refused to meet together to honour their dead.'[28] De Valera hinted at the contention that surrounded the memorial when he made a point of referring to the modest nature of the monument and said that 'the time to raise a proud national monument to the work that was here begun and to those who inspired and participated in it has not yet come. Such a monument can be raised only when the work is triumphantly completed.'[29] The leader of the opposition and of the Cumann na nGaedheal party, W. T. Cosgrave, argued that what he and others had fought for in 1916 had not yet been achieved and that the unveiling of the memorial was somewhat premature. As a report in the *Irish Press* put it, 'It is not possible to hide these national limitations today or to cover them with a veil lifted from the bronze statue of Cuchulainn.'[30] More militant republicans staged a more overt protest when they paraded through the streets of the capital in an attempt to rival the official military display. This took place later in the same afternoon when an estimated thousand people marched in formation to the republican plot at Glasnevin Cemetery where the Chief of Staff of the IRA, Mr Maurice Twomey, delivered an oration.

Plate 6.3 **1916 memorial, General Post Office**
Erected 1935, sculpted by Albert Power
Bord Fáilte

Once in place in the public hall of the GPO, the Cuchulainn monument became an important site of national memory in a manner almost akin to a national war memorial. The association of the legendary Celtic warrior from a previous golden age with the ideals of those who had died in the 1916 Rising forged a link between past and present and contributed to the cultivation of a monumental landscape that marked Ireland's emergence from beneath the shadow of colonial rule. The erection of the memorial further reinforced the symbolism of the GPO as the focal point of the rebellion, and while visitors flocked to see the figure of Cuchulainn, the space around the Post Office was regularly used for commemorative parades and nationalist processions. The memorial was just one of a range of sculptural initiatives launched by the Free State administration in an attempt to forge national identity and demonstrate the distinctive nature of the Irish national character.

New heroes for old spaces: memorialising Markievicz and Heuston

One of the first monuments commissioned while de Valera was President was dedicated to a female participant in the 1916 rising, Constance Gore-Booth, Countess de Markievicz, TD. The monument had been initially commissioned in the 1920s by a group of republicans some of whom were in government when the monument was unveiled, and took the form of a bust sculpted by Albert Power.[31] It was made of Irish limestone and stood six and a half foot high, with the inscription; 'Constance Gore-Booth, Countess de Markievicz, T.D. 1868–1927, Major in the Irish Citizen Army.' The words, 'A valiant woman who fought for Ireland in St. Stephen's Green, 1916', appear on the right.[32] The monument was positioned significantly in St Stephen's Green, where she had served as second in command to Michael Mallin during the 1916 Rising when a contingent of the Irish Citizen Army attempted to take control of the area. It was unveiled in a ceremony in the Green on 3 July 1932. The chair of the memorial committee, Senator L. O'Neill noted in his unveiling speech that:

> it was not necessary to raise a statue or a monument to Countess Markievicz. Her memory would live forever in the heart of the Gael. She left them five years ago, but many years before that she put aside the wealth and position that might have been hers and chose the hard road that they must travel who truly loved the poor and the lowly.[33]

The unveiling of a portrait statue of Sean Heuston, one of the leaders of the 1916 rebellion, took place on 5 December 1943. His statue was erected in the grounds of the People's Gardens in the Phoenix Park on the outskirts of the city, but close to Kingsbridge Train Station, which was later renamed in his honour.[34] A memorial committee had approached the government in 1939 with the proposal. The principal architect of the OPW, T. J. Byrne, advised the government to contact the Metropolitan School of Art in search of a sculptor. Laurence Campbell was selected and his portrait statue of the hero of 1916 was unveiled to a crowd of 2,000 people. Heuston was represented in Irish limestone and dressed in military uniform and an inscription was attached which noted; 'Captain of the 1st battalion Dublin brigade Irish Volunteers, Officer commanding the Mendacity Institute in Easter Week, Executed Kilmainham 8th May 1916'.

Plate 6.4 **Constance Gore-Booth, Countess Markiewicz, St Stephens Green**

Plate 6.5 **Sean Heuston, Phoenix Park**

Dublin Brigade Memorial, Custom House

The commemoration of the fortieth anniversary of the 1916 Rising in 1956 provided the impetus for the erection of a number of other public memorials, among them the Custom House memorial. This was dedicated to those members of the Irish Republican Brotherhood (IRB) who had died in the War of Independence when the Custom House came under attack on 25 May 1921. Sculpted by the Breton nationalist, Yann Renard-Goulet, the monument represented Ireland in the guise of woman holding a sword and stretching out a hand to the dying soldier at her side while at the same time looking with determination to the future and to those who would carry on the fight. The dying soldier, meanwhile, lies with a broken manacle on one wrist depicting the breaking of the bonds of servitude while, from the rock on which the figure of Ireland stands, an oak tree juts out which has burst into leaf signifying the revival of another generation (see plate 6.6). [35]

The entire project had been initiated by the Dublin Brigade Memorial Committee and was formally unveiled on 20 May 1956. The product of a

ten-year gestation, paid for by the contributions from thousands of citizens of Dublin, the memorial was intended to serve as a tribute not only to the dead but also to their comrades who survived and who had lived to see the outcome of their actions. Once again it brought together a large gathering of politicians and interest groups. The event, which was presided over by Oscar Traynor who had commanded the Dublin Brigade at the time of the Custom House operation, was also marked by a procession from Dublin Castle to the Custom House. In his unveiling speech, President O'Kelly said:

Plate 6.6 **Dublin Brigade Memorial, Custom House**
Erected in 1956, sculpted by Yann Renard-Goulet

We could now look for the first time on a worthy memorial of one of the most famous events of that period [1916–21], the burning of the Custom House. Next to Dublin Castle it had been the most important centre of British administration in Ireland. Like the castle it was a symbol of a claim to rule Ireland by external force, which was repudiated by the Irish people. The destruction of the local government boards records and those of Customs and Inland Revenue offices, it was recognised would go far towards disorganising the remaining British efforts to conduct government of Ireland in civil affairs . . . but even more important than this very political reason, was the desire to demonstrate . . . to all the world that the vigour and resilience of the Irish struggle for freedom had not been weakened by the ruthless pressures of its opponents.[36]

1916 memorial, Arbour Hill

Also in 1956, a site on the north side of the city at Arbour Hill was given over to the formal commemoration of the dead of the 1916 Rising. A stained glass window in memory of all who had died from 1916 to 1923 had been erected in the nearby church of the Sacred Heart four years earlier. By 1948 the graves had become a place of annual pilgrimage and the idea of erecting a suitable memorial was raised again in the Dáil in 1954, when:

> Mr Brady asked the Minister for Defence if he will state what progress has been made in regard to the proposed memorial to the executed 1916 leaders at Arbour Hill; further, whether it is proposed, in the meantime, to mark the grave with a simple slab giving the names of the executed men.[37]

In his reply, the Minister, General MacEoin, informed the Dáil that:

> Plans for a memorial scheme at Arbour Hill Detention Barracks, to commemorate the 1916 leaders who are buried in the prison yard, have been prepared. I am having them examined, and I propose to submit them to the government for consideration when I have the examination completed. Pending the consideration of the general scheme, it is not proposed to proceed with the marking of the graves.[48]

The memorial was completed in 1956, comprising three paved terraces with a curved wall backdrop on which the sculptor Michael Biggs inscribed the Proclamation of the Republic in both Irish and English, while the names of the leaders were carved on the kerb surrounding the burial plot. On 24 April 1956, as part of the annual commemorative ceremonies, the President, Sean T. O'Kelly, along with the Taoiseach, Eamon de Valera, and several members of the government and the judiciary, the Lord Mayor of Dublin, members of the Oireachtas, veterans of the Easter Rising and relatives of the 1916 leaders, attended a special memorial mass celebrated in the church of the Sacred Heart at Arbour Hill. After the mass a procession took place to the 1916 Memorial Plot where the President unveiled the large limestone plaque. The memorial subsequently became a focal point of the annual 1916 commemorations, not least in 1966, something of a watershed year which marked the fiftieth anniversary of the Rising.[39]

Remembering 1848: Thomas Davis and O'Donovan Rossa

Alone of all the Irish poets
You took the difficult weapon up
Not to carve beauty out of stone
But to set the sudden crop

All the young harvesters at that work
Saw thistle ragwort flourishing
Rejoice then that you left the field
In the red-furrowed spring.[40]

September 1945 marked the centenary of the death of the Irish poet and patriot, Thomas Davis, and provided the catalyst for a display of Irish nationalism on the streets of the capital when the foundation stone of a statue dedicated to him was laid in College Green. Initially, the then principal architect of the OPW, Raymond McGrath, was requested to prepare a design for a memorial and he chose a site in St Stephen's Green where a central fountain was to be placed at the spot where the statue dedicated to George II had been.[41] His design incorporated curved pavilions in neo-classical style which were to flank the fountain on either side (plate 6.5). Later dubbed 'Nuremberg in Dublin' McGrath's plan was never executed. Rather, in a gesture of some significance, a site on College Green formerly the location of the statue dedicated to William III was allocated for the Davis monument.

The foundation stone for this monument, close to the university that Davis had attended, was laid in September 1945 as part of the week-long centenary celebrations to mark his death. The week began with a parade of over 10,000 people through the streets of the capital.[42] At the head of the procession was the Lord Mayor, Alderman P.S. Doyle TD, 'who led the march from Parnell Square through O'Connell Street, Westmoreland Street and Dame Street to the City Hall. Members of the city authority, Dublin Corporation followed him in their ceremonial robes, while units of the local defence force lined the route along with large crowds.'[43]

> After the final line-up the Lord Mayor, Mr. de Valera and members of the Government, the City Councillors, members of the Mallow U.D.C., descendants of some of the Young Irelanders, prominent figures in the Irish political parties and members of the Dail and Senate went out upon a platform in front

Plate 6.7 **McGrath's design for the Thomas Davis memorial**
Irish Architectural Archive, with thanks to Jenny O'Donovan

of the City Hall, where the Lord Mayor placed a wreath on the statue of Thomas Davis, which had been removed outside from the City Hall for the celebrations.[44]

Three days later on 12 September, the foundation stone of the Davis memorial was laid by the President, Sean T. O'Kelly:

Here, where Thomas Davis often walked as a student and in the short years of his public life, I have placed this tablet to mark the site of a statue to be erected to his memory . . . Davis loved statues; those embodiments in stone of people's history . . . Thomas Davis was one of a line of great men who stretched backwards through Grattan to the days of our former independence and forwards through Mitchel, Parnell and Childers to the present independence of this part of Ireland . . . When we remember that Davis belonged to the minority, which at that time possessed practically all power and patronage in the nation his advocacy of an equality as Irishmen of those of all creeds is seen in its real greatness. He urged, as Tone had urged before him, that all differences be sunk in the common name of Irishmen serving the common cause of Irish freedom. Such a union of Catholic and Protestant, Williamite and Jacobite, Cromwellian and Milesian, he saw as the source of benefits to our nation such as patriots had long dreamt of.[45]

It was, however, another twenty years before a statue, designed by Edward Delaney, was unveiled in 1966, in conjunction with the fiftieth anniversary of the 1916 Rising.[46] Delaney's design was made up of two main components, one a ten-foot bronze statue of Davis atop a nine-foot high pedestal made of granite quarried in the Dublin Mountains. The other part of Delaney's design centred around a fountain pool which was eleven foot in diameter and arranged at two levels with water overflowing from the upper level into the surrounding pool. The fountain was surrounded by bronze figures representing the heralds of the four Irish provinces. Each in turn stood on an ornamental framework surrounded by six granite tablets bearing bronze reliefs illustrating the poetry of Davis, namely; 'The Penal Days', 'Wolfe Tone's Grave', 'The Burial', 'We Must Not Fail', 'A Nation Once Again', and 'The Famine' (plate 6.8).[47]

The monument was ready for unveiling in April 1966 by President Eamon de Valera. He mentioned in his unveiling speech that the occasion of the 1916 commemoration was a suitable time for the unveiling of a monument dedicated to Davis given that 'the men of 1916 were his spiritual children'.[48] Moreover, the fact that the statue was located on the site of the monument formerly dedicated to King William III and its proximity to Ireland's former Parliament House, was not lost on the participants. The president went on to observe that:

> Davis, the Young Irelanders and O'Connell had tried to win that free parliament again, but Davis wanted not a parliament of the aristocracy but a parliament for the common people, a free parliament for all Ireland. He, like Tone and Emmet, and the men of 1916, wished to unite all the people of Ireland and to abolish the memory of past dissension.[49]

After the monument had been unveiled, the fountains were put to work and the trumpeters and drummers from the Army Number One Band sounded a salute. To the accompaniment of the band, verses in English and Irish from 'A nation once again' were sung by a choir of school children, while the ceremony concluded with the National Anthem. Also in attendance at the ceremony were members of the government, Dáil and Senate, Council of State

Plate 6.8 **Thomas Davis, College Green**
Erected 1966, sculpted by Edward Delaney

and the Judiciary, the Chancellor of Dublin University and Provost of Trinity College, the President of University College, along with relatives of Davis and other Young Irelanders. The anniversary also saw the unveiling of a stone column in St Stephen's Green erected in honour of former members of Fianna Éireann. Composed of limestone it was designed by A. J. Breen.

The monument dedicated to Jeremiah O'Donovan Rossa, the Fenian leader and poet, was unveiled with much public display on 6 June 1954.[50] The occasion was an overtly political one that brought together five battalions of the old IRA, members of Cumann na mBan, the Irish Citizen Army, Fianna Éireann and the Dublin Brigade, along with several key members of the political administration in power at the time, most notably the President, Sean T. O'Kelly and the Taoiseach, Eamon de Valera. The monument was put in place just inside the entrance of St Stephen's Green and took the form of an uncut piece of Wicklow granite, designed to underline the unbreakable spirit of the poet. A bronze relief of O'Donovan Rossa by the sculptor Seamus Murphy was set into it and inscribed with the words, 'The Gaels will not forget you forever', from Patrick Pearse's funeral oration (plate 6.9).[51] As the Taoiseach pointed out in his unveiling speech, the monument was symbolic of the man it was placed there to honour and would serve:

> to remind many generations of an indomitable fighter given to Ireland out of west Cork, cradle of many gallant Irish patriots. In particular I would have it remind this and coming generations that here indeed was one dauntless man who, in spite of threats, bribes, imprisonments and hunger and in spite of inhuman tortures inflicted on him by the courts and minions of an alien government, never wavered, never compromised but was faithful to his last hour in the cause he had early sworn to serve: the cause of the Irish Republic.[52]

Plate 6.9 **O'Donovan Rossa, St Stephen's Green**
Erected in 1954, sculpted by Seamus Murphy

The monument, erected by the O'Donovan Rossa memorial committee and sanctioned by the OPW, effectively served as a tangible reminder of the ideals that he had set out to achieve of making 'a free and Gaelic Ireland', and also of the fact that these ideals had not yet been achieved. The President observed that:

> The monument might also serve as a reminder to us that the object Rossa and his companions set out to achieve – the making of a free and Gaelic Ireland – should not have been accomplished until every acre of the soil of our land should have been delivered from foreign rule and the language of ancient Ireland had been restored to its rightful place. We can best honour the memory of O'Donovan Rossa by doing our duty, each one of us, to make the realisation of these objectives possible.[53]

Commemorating the fiftieth anniversary of the 1916 Rising

Some weeks after the destruction of Nelson's Pillar in March 1966, the Golden Jubilee of the 1916 Rising took place and once again O'Connell Street played an important role in the annual spectacle of commemoration and celebration. The purpose of the two-week long, countrywide commemorative celebration was to 'honour those who took part in it and to emphasise its importance as a decisive event in our history'.[54] Newspapers issued special souvenir numbers that recounted details of the Rising, the heroic figures that took part in it and the key events of its course. On Easter Sunday 1966, thousands came out onto the streets of the capital to witness the parade of six hundred veterans of the Rising, many of whom had been members of the Dublin garrisons.[55] The ceremony continued with a solemn reading of the proclamation of the Republic that rang out from loudspeakers, followed by a 21-gun salute and military march. The route took the participants from O'Connell Street to many of the key places associated with the Rising, among them the Four Courts, the Mendicity Institute, Jacob's Factory, Boland's Bakery, Mount Street Bridge and the South Dublin Union, before concluding with a rendition of the national anthem by the Army Number One Band. The day closed with the premiere of a film specially commissioned as part of the commemoration, a retrospective look at the events of the Rising which was then released for distribution throughout the country. The rest of the week continued in much the same vein with the Taoiseach and President appearing at several official commemorative functions. Special ceremonies took place at provincial centres all around the country, while a number of significant statues were unveiled in the capital, along with the formal opening of the Garden of Remembrance, just north of O'Connell Street at Parnell Square.

Opening the Garden of Remembrance, Parnell Square

*In the darkness of despair we saw a vision. We lit the light of hope and it was not
    extinguished*

*In the desert of discouragement we saw a vision. We planted the tree of valour and
    it blossomed.*

*In the winter of bondage we saw a vision. We melted the snow of lethargy and the
    river of resurrection flowed from it. We sent our vision aswim like a swan on the
    river. The vision became a reality. Winter became summer. Bondage became
    freedom and this we left to you as your inheritance. O generations of freedom
    remember us, the generations of the vision.*

(Inscription at the Garden of Remembrance, Parnell Square)

The opening of the garden marked the culmination of a project that can be traced back to September 1935 when the Dublin Brigade Council of the Old IRA suggested to Government that a site on the northern part of the Rotunda Gardens be converted into a memorial garden. The location, which held particular symbolic significance as it marked the site where Óglaigh na nÉireann was founded in 1913 and was also where prisoners of 1916 had been held during Easter Week, was acquired from the Governors of the Rotunda Hospital in October 1939 at a cost of £2,000. In March 1940 a design competition was launched but consideration of entries postponed owing to the outbreak of the First World War. Consequently, the winning entry, submitted by the architect Daithí P. Hanley, was not announced until 20 August 1946.

Hanley's design centred on a sunken garden in the form of a cross to symbolise the dead and enclosing a pool. In design, the plan drew heavily on Ireland's Celtic past and made use of much religious iconography. The floor of the pool was given a mosaic pattern of blue-green waves into which were set various Celtic weapons intended to symbolise the ancient custom of throwing weapons into water on the cessation of hostilities. At one end of the pool Hanley left a space for a sculptured monument that was backed by a curved white marble wall. Inserted into the railings at various points around the sunken pool were copies of artefacts held in the National Museum in Dublin, including the Brian Boru harp, the Loughnashade trumpet with the Cross of Cloyne set above it and the Ballinderry sword pointed downwards to symbolise peace. Hanley also provided a striking entrance for the garden, the centrepiece of which was a set of gates, 50 feet wide, upon which was placed the title 'Gairdín Cuimhneacháin' together with a bronze replica of the processional cross of Clogher.

The architect also incorporated into his design a proposal for a sculpture in bronze representing Éire to be erected on the pedestal at the top of the sunken pool. This figure was:

> guarded by four warriors of the provinces with a background of patriots in bas-relief . . . The statue is symbolic of the inspiration and idealism for which the patriots lived and died. On the walls of the sunken garden are the names of patriots on sculptured county memorials . . . Niches containing busts of patriots could be added later if desired.[56]

The design also provided for seating in the garden, as well as 'portable flower-boxes of geraniums and tulips and blossoms of climbing aubrietia and rock plants on the retaining wall. This would help to make it a pleasantly sheltered place to walk in contemplation or to sit and rest beside the reflecting pool. Ireland's youth should be inspired by the names and sagas of the past.'[57] Hanley based his design on the premise that there should be a degree of intimacy between the memorial garden and passers by, 'Its monumental features should be easily seen and recognised. It should also be "insulated" from its varied architectural surroundings. People sitting in the garden and viewing it should feel secluded. It should inspire people with a feeling of respect for the patriot dead and yet be sufficiently light in treatment to be used as a small quiet garden in which to sit and rest (plate 6.10).'[58] He also took into account the contemporary preference for marches on national holidays:

> On national holidays a march past would pass along the route as at present, passing O'Connell Street and Parnell Square, East, and so pass the gates of the proposed memorial garden. A wreath could be laid on the stone platform at the front of the monument. A guard of honour might stand on the crescent shaped terrace behind the statue, overlooking the garden. A volley could be fired from here. Distinguished visitors might be provided with seats on the lawn.[59]

Almost immediately after the winning design was announced, controversy arose regarding the chosen site on Parnell Square. As a commentator in the *Irish Builder* put it:

> many architects and town planners, including the winner of the competition are not in favour of this proposal to build a memorial in the Rotunda gardens, as they feel that the site is not adequate for a national memorial worthy of those whom it is intended to honour. While it is felt that the memorial would be quite suitable for a period of history such as the 1916 Rising, a national memorial deserves something more inspiring and on a bigger scale than anything that could be achieved in such restricted and secluded surroundings.[60]

This was echoed shortly after in the Dáil where the issue was discussed in 1946. Many deputies argued that the small area of ground available made the site inadequate.[61] By March 1949 the issue had not been resolved and was complicated by the fact that plans for a neo-natal unit adjacent to the Rotunda Hospital in the grounds of Parnell Square also found support in the Dáil. In the heated exchanges that followed, Sean McEntee, TD, made clear his feelings on the matter, declaring 'So that is the way the republicans are going to deal

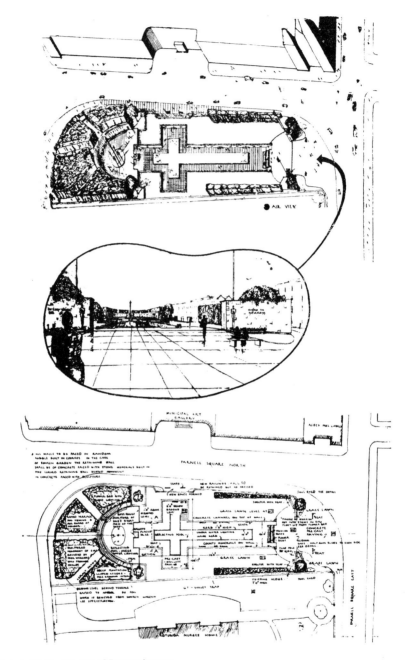

Plate 6.10 **The Garden of Remembrance**

From the *Irish Builder*, 7 September 1946

with the Garden of Remembrance?'[62] The Minister responsible replied: 'The Garden of Remembrance, according to my file, was first thought of in the year 1935. It took all that time for the last Government to think of the matter . . . The site is not suitable at all.'[63] The debate continued to rage in the Dáil, where in July 1949 Mr C. Lehane, TD, asked the Minister for Health, Noel Browne, whether he was aware that the plot of ground intended for use in connection with a neo-natal clinic, was already earmarked as 'the site for a national memorial to all those who laid down their lives in the fight to establish an Irish Republic'.[64] He went on ask the Minister, 'if he will state whether he has taken over this site permanently or merely as a temporary expedient'.[65] In reply, Browne stated his awareness of the position and that:

> it was only in consideration of the urgency of taking action to reduce infant mortality in Dublin that I agreed to recommend that the memorial plot should be made available as a site for a temporary infant welfare unit at the Rotunda Hospital. This is a purely temporary arrangement and intended to meet the position only while the hospital authorities are developing proposals for a permanent unit which will not entail encroachment on the memorial site . . . So far as I am aware there is no question of abandoning the proposed scheme for the provision of a memorial park in that area.[66]

It was not until 1957 that the issue was finally resolved when the infant unit was transferred to the grounds of the Rotunda Hospital and the way was paved for the creation of the memorial garden. Work began in 1964 and it was completed in time for the fiftieth anniversary commemoration of the Rising.

The Garden of Remembrance was finally opened to the public in a carefully choreographed ceremony that brought together the forces of Church and State. At the outset the architect presented the President with the key to the garden, a three-times enlarged copy of the oldest known Irish key that had been found at an excavation near the Hill of Tara. De Valera then opened the gates to the sound of a fanfare from the army trumpeters gathered close by. Both the President and the Taoiseach, Sean Lemass, then took up their positions in the centre of the sculpture platform. To their right were gathered members of the Government, ex-Ministers, the Lord Mayor and high-ranking army officers, while the Archbishop of Dublin, members of the judiciary, various public representatives and members of the Labour Party were clustered to the left, with members of the diplomatic corps situated just below them. The ceremony began with a blessing by the Archbishop of Dublin, John Charles McQuaid, after which de Valera delivered the opening address. He began by thanking assembled Church leaders for leading prayers and then went on to draw particular attention to the appropriate nature of the site chosen for the national memorial, one that was so closely associated with the independence struggle:

> The site was beside the old rink in which the Irish volunteers were founded on November 25th 1913. Not far away was the General Post Office where Padraig Pearse proclaimed the Republic in 1916. At the southeast corner of the Square were the Rotunda buildings, where many memorable meetings were held. It was there the first Oireachtas of the Gaelic League was held. It was also the scene of the meetings of the Land Leaguers and of the Irish Parliamentary Party . . . On the north side of the square were the old headquarters of the Gaelic League . . . On the west side of the square were the houses where the Irish Volunteers executive met and also the headquarters of the Irish Citizen Army . . . the site was in every way eminently suitable.[67]

De Valera also alluded to the purpose of the garden and spoke of his desire that it would remind people of the sacrifices, struggle and suffering endured over the centuries to secure independence. He concluded by expressing the hope that visitors to the garden would remember 'not only the leaders of the Rising but also the nameless ones, the unknown soldiers, the men and women in the ranks who bear the burden of every battle'.[68] He spoke in Irish when formally declaring the garden open, at which point the National flag was hoisted to full mast. Towards the end of the ceremony the 'The Last Post' was sounded and De Valera laid a wreath on behalf of the people of Ireland.

A key element of Hanley's plan for the Garden of Remembrance was for a central sculpture which would symbolise the national struggle and occupy the circular plinth at the western end of the garden. Although he had incorporated plans for this into his original design, the space was left vacant until 1971 when the Dublin sculptor, Oisín Kelly, was approached and requested to write a report on the proposed sculpture. In this report Kelly stated his opinion that 'I cannot over-emphasise, that the sculpture is not an ornament in the modern and incorrect sense of the word as something added to increase the beauty of an object . . . the sculpture signifies the purpose, the serious and unique purpose of this garden and deserves a site worthy of that purpose.'[69] He went on to lament the choice of site, arguing that:

I doubt if any architectural device can ever cloak the fact that what was once a pleasure garden of a residential square will be variously allotted to a maternity hospital, a nurses' home, a theatre, a dance hall and a garden of remembrance to a nation's heroes. In this site, the Garden of Remembrance can never be more than a part of a whole . . . we must have a site where it does not compete with so many discordant voices.[70]

He suggested instead a site in the Phoenix Park 'where the necessary scale and space are available. Another site might be, for example, Fairview Park, or further along the new coast road to Howth'.[71] Despite these reservations about the Parnell Square site, which was out of his hands anyway, Kelly proceeded to design a memorial sculpture. He rejected Hanley's initial proposal for the figure of Éire and her four warriors, arguing that such a proposal constituted, 'an insoluble historical, psychological and sculptural problem, and should be abandoned'.[72] He believed that a national iconography did not exist in Ireland, rather that:

most memorials are a conglomeration of foreign elements, Irish in nomen-clature and detail but alien in spirit, elements from Norman heraldry and elements from English Victorian sentimentality such as the Albert memorial. The most successful national memorial has been the death of Cuchulainn in the GPO, although its scale is too small for its position.[73]

He eventually put forward a design that drew its inspiration from both the legendary Irish saga of the Children of Lir and the poem 'Easter 1916' by W. B. Yeats with its central idea that men at certain moments in history are 'transformed utterly'. In order to illustrate this theme, Kelly made use of the Children of Lir story and sought to represent the agony of transformation in his sculpture. The use of the swan motif meanwhile, a 'generally accepted image of resurgence, triumph and perfection, with undertones of regal sadness and isolation' more than adequately served his purpose of creating a memorial, 'which does not attempt to bully my countrymen into having splendid thoughts and noble feelings, but rather one whose message was implicit, a hint rather than a shout'.[74]

The 'Children of Lir' sculptural group was eventually unveiled on 11 July 1971, on the day that marked the fiftieth anniversary of the 1921 truce between British and Irish forces (plates 6.11 and 6.12). It was formally dedicated by President de Valera to the memory of all those who gave their lives in the cause

Plates 6.11 and 6.12 **The Children of Lir sculpture and Garden of Remembrance**

Unveiled in 1971, sculpted by Oisín Kelly

of Irish freedom. The occasion was marked by a controversial speech delivered by the Taoiseach, Jack Lynch, in which he stated his belief that, 'We have justi-fied the struggle for our freedom and fulfilled many of the hopes and aims of those who fought for our freedom.'[75] He went on to urge the British Government to declare its interest in encouraging the unity of Ireland by agreement. Such an 'historic step', he argued, 'would forward the work begun fifty years ago, when Britain and the Irish nation agreed to a truce'.[76]

The well-choreographed spectacle that accompanied the formal opening of both the memorial garden in 1966 and the sculptural focal point in 1971 marked the culmination of a lengthy gestation. The garden provided a very public platform for the commemoration of those who had been killed in the Independence struggle and as such contributed to the symbolic expression of nationhood. The emphasis on religious iconography and ancient Celtic motifs in Hanley's design, coupled with Kelly's bronze sculpture, served not only to commemorate the dead effectively but also to draw a parallel between them and the ancient warriors of the heroic, Celtic and, significantly, pre-colonial past. This was reinforced during the unveiling ceremony and in the speeches that were delivered not only at the Garden of Remembrance but also during the unveiling of many other monuments in the years leading up to 1966 and especially during the commemorative events associated with the Golden Jubilee celebrations. In the aftermath of the flurry of activity that accompanied the Jubilee, Dublin's monumental landscape gradually began to change. Thereafter, while public thoroughfares continued to act as sites for public monuments, these increasingly became apolitical sites of public art.

The Irish National War Memorial, Islandbridge

It was perhaps inevitable that after centuries of British rule, the Free State should begin its independent life by asserting itself as a Catholic democracy with an explicit emphasis on its Gaelic character. Public monuments were a most visible expression of this trend. Successive administrations were adept at using monuments, and the choreographed ritual that went with their unveiling, to best advantage. Monuments were erected dedicated to the heroes of the rebellion as part of a strong desire to make real the struggle that the country had come through. It is significant, however, that the 50,000 Irishmen who had died in the Great War fighting on the side of England and the allies were com-memorated in a much more low-key manner. The National War Memorial at

Islandbridge in the west of the city was completed in 1939 and the park was dedicated simply 'To the memory of the 49,400 Irishmen who gave their lives in the Great War, 1914–18'.[77]

As the Great War drew to a close, the question of commemorating those who had been killed came to fore, influenced no doubt by the flurry of monumental activity in Britain.[78] The project to erect the Irish national war memorial can be traced to July 1919, when:

> a representative meeting of upwards of one hundred Irishmen, hailing from North, South, East and West, met and resolved that a permanent memorial should be erected to commemorate the Irish officers and men who had fallen in the Great War, and an executive Committee was appointed to effect the resolution.[79]

An appeal was made for public subscriptions and almost £50,000 was subscribed. The memorial committee then set about the task of finding a scheme which, according to legal obligations, had to be public, visible and monumental.[80] Initially a scheme was proposed which provided for the building of a central home called 'The Great War memorial Home', to be equipped to provide board, lodging and recreation for soldiers and sailors passing through Dublin. It was also proposed that a Record Room should be set apart 'to contain a record of the names of the Irish Officers and men of all the services who were killed in the Great War'.[81] This proposal was rejected by the committee, however, which opted instead for what was to become known as 'the Merrion Square scheme'. This provided for the opening of Merrion Square to the public and for the erection in its grounds of a memorial centrepiece and gates.[82] While the scheme was approved by the Courts in March 1926, and Dublin Corporation agreed to take over and maintain the square as a public park, a year later the Irish legislature rejected the plan.[83] Kevin O'Higgins, a Minister in the Free State government, argued that the sacrifices of the Great War should be set against those of the Easter Week. He said that 'a war memorial set beneath the windows of government buildings would mislead strangers and would give a wrong twist to the origins of the state'.[84]

The scheme was eventually settled in 1929 when the committee once again approached the government. The members highlighted the fact that they had been unable to find a site on private ground and asked the government for a position in the Phoenix Park. President Cosgrave suggested a site at Islandbridge and presented his scheme for a National Memorial Park, to be laid out on the bank of the Liffey, at a site known as Longmeadows. The committee agreed to this proposal and in 1930 Sir Edwin Lutyens was commissioned to carry out the designs.[85] Building work began on the site in December 1931 and proceeded under the de Valera administration after 1932.[86]

> The design of the memorial is by Sir Edwin Lutyens, R.A. and is a notable example of the simple dignity so characteristic of this great architect's style. The memorial is designed as a Garden, and Sir Edwin, during his many visits, always took pains to impress this aspect upon those carrying out the work. The simple war stone in the centre of the lawn, the magnificent cross standing at the head of an imposing semicircular flight of steps, the book-rooms with their graceful columns and deep cornices, the fine fountains with their broad basins, the pergolas, the Gate-piers, the wall-piers, the niches, all these arrest the attention and compel admiration.[87]

Lutyens's design for the war memorial incorporated a war stone of Irish granite, identical in size and shape to war stones erected in honour of war dead all over the world. It was flanked on either side by fountain basins with obelisks in their centre, symbolising two candles on either side of an altar. A cross, with truncated arms, was aligned with the war stone and approached by a flight of steps.[88] Granite pergolas were erected at either end of the lawn in order to divide the centrepiece from two circular terraced rose-gardens, each containing more than 4,000 rose trees. At the ends of each of the pergolas, four pavilions, or book rooms, were built, into which copies of illuminated volumes containing the names and recorded service of the 49,400 Irishmen who were killed in the Great War were placed (plates 6.13 and 6.14).[89]

Between 1933 and 1938 the memorial garden was laid out in two phases. The first involved the laying out of the parkway stretching from Islandbridge to Chapelizod, and the second the laying out of the garden of remembrance, a 20-acre plot within the park. This work was financed by a donation of £50,000 from the Irish government and a similar sum was donated from the war memorial fund. The park took two years to complete using a labour force of 300 men. The agreement to proceed with the remembrance garden was formalised between the government and the war memorial committee on 12 December 1933. At noon of that same day work was begun. Granite was manually extracted from Ballyknockan, County Wicklow and Barnacullia, County Dublin:

Our Irish granite quarries are not equipped with mechanical tools, and we cannot but admire the skill of the men who with hammer and chisel carved out these beautiful columns to be seen in the book rooms and elsewhere, and formed the intricate mouldings included in the design of every one of the buildings.[90]

The work on the garden was completed by 1939 and the park was simply dedicated:

To the memory of the 49,400 Irishmen
who gave their lives in the Great War,
1914–18.

Provision was also made for the erection of crosses on the battlefields of Europe to commemorate the Irish soldiers who had fallen there. The sixteenth Irish division had taken part in the Battle of the Somme in September 1916. On 3 September they captured Guillemont and on 9 September they stormed the village of Ginchy. In the course of these battles they lost 263 officers and 4,091 other ranks; a total of 4,354 died. By 1925 many of the divisions and armies had erected memorials on the battlegrounds, hence it was felt that some sort of memorial should be erected there for the Irish soldiers. The Divisional Commander of the Irish, Major General Sir William Hickie, arranged for the erection of a Celtic cross in wood, 13 foot high and made from a baulk of elm which came from a ruined French farmhouse. It was erected between the battlefields of Guillemont and Ginchy, and although this territory was lost to the Germans in their attack of March 1918, it remained there until 1926 when it was replaced by a similar cross made of Irish granite and erected in the graveyard of the church of Guillemont as a French law forbade the erection of crucifixes or crosses outside the precincts of a church. It was unveiled by Marechal Joffre and blessed by the Bishop of Amiens in May 1926 in the presence of a large body of Irishmen and women. The original wooden cross was taken back to Ireland and from 1926 to 1939 was erected annually in the Phoenix Park on the ground just east of the Wellington monument where it served as a cenotaph during Remembrance Day ceremonies. It was subsequently placed in the northeast house of the newly completed memorial where it remains today.[91]

The formal opening of the national war memorial in Dublin was, however, delayed, owing to the outbreak of the Second World War.[92] In the intervening years the gardens fell into a state of disrepair which necessitated a major

Plate 6.13 **Irish National War Memorial, Islandbridge**
Designed by Edwin Lutyens, 1930–1940, aerial view, OPW

Plate 6.14 **Irish National War Memorial**
Rose-garden and book-rooms, OPW

restoration programme. This was carried out in the late 1980s under the auspices of the OPW.[93] The restoration was completed in 1988 and the park opened and blessed by Church leaders. Government representatives were missing from the opening and it was not until April 1995 that it was given an official government opening.[94]

Conclusion

The capacity of the seemingly innocuous public statue to engage with popular public opinion, to shape ideals and political values and to contribute to the nation-building process, was made patently clear in Dublin in the decades after independence. The various State-sanctioned monumental initiatives created a visual landscape that served to fix in the minds of citizens the memory of people and events that marked out the path to political independence. As Bhreathnach-Lynch writes: 'One important reason to rush to create "new" heroes and revive "old" ones lies in the need a newly independent nation has to establish quickly its own sense of national identity . . . The promotion of new heroes suggested not only a positive, affirming self-image but also implied that another such age would emerge in the near future.'[95]

The bitter and contentious Civil War that followed the emergence of the Irish Free State, however, meant that commemorative practices in Dublin and more widely across Ireland were lacking in direction from central government. As Fitzpatrick points out:

> The Irish Free State emerged from four bitter conflicts, between Irish and English, Catholics and Protestants, 'constitutional nationalists' and 'separatists', and finally 'Treatyites' and republicans. The rulers of the new state represented a sub-group within Catholic and nationalist Ireland, their triumph being bitterly resented by many Irish protestants as well as by Catholic republicans and unreformed constitutionalists.[96]

This political discord inevitably raised problems for commemorating the past and often flared up during unveiling ceremonies, such as occurred at the Civil War commemoration ceremony and at the unveiling of the 1916 monument in the General Post Office. Nevertheless, a host of figures drawn from the disparate strands of Irish nationalist politics did find their way onto the streets and green spaces of the capital after 1922. This 'scripting of national memory' stands in marked contrast to the de-commemorative practices which also prevailed after 1922. These saw older icons of empires unceremoniously bombed from their pedestals or sold by the state to foreign interests, among them the statues of kings William III, George I and George II, along with monuments dedicated to Lord Gough, the Earl of Carlisle and the Duke of Eglinton and Winton. In 1948 the statue of Queen Victoria was removed from the grounds of Leinster House in a symbolic gesture that coincided with Ireland's departure from the Commonwealth. It is to eradication of this legacy of empire that the next chapter turns.

CHAPTER SEVEN

Removing icons of empire: De-commemoration and the cultural landscape

Destroying icons of empire

Nelson, Queen Victoria and other British statues are ancient monuments, trophies left behind by a civilisation which has lost the eight centuries battle. The hand that touches one of them is the hand of an ignoramus and a vandal.[1]

The decades after 1922 were also marked by the gradual and frequently illegal removal from the urban landscape of virtually all the monumental symbols of the British empire. Shortly after the foundation of the state the obelisk at Oldbridge commemorating King William III's victory at the Battle of the Boyne was bombed by a freelance party of the Irish Army based in Drogheda, who drove out to the Boyne and destroyed it with landmines. The incident led *The Irish Times* to remark:

> Is history to be for Irishmen nothing save perpetual irritant? Can it not teach them something of the process by which they have become what they are now, of the quarrels they have survived, of the slow blendings and absorptions that have united Celt and Norman, Ironside and Hessian, of time's, transformations and severances? All these things are written in many books, but they are written most vividly on the monuments of our land, and these monuments, whether the things that they recall are gracious or painful, ought to be held among our most cherished possessions. That is the civilised practice of other lands . . . In its essence the outrage at Oldbridge is not less foolish and wicked than if landmines had been exploded among the guarded pillars of Mellifont or Cong. Every Irishman is the poorer by any deed which weakens Ireland's links with her past.[2]

The blowing up of the obelisk was merely a signal of things to come and the sentiments expressed in *The Irish Times* were to be repeated on a number of occasions throughout the decades after 1922. It should be borne in mind, however, that support for such actions was not uniform and neither was it government policy to remove statues. The much-maligned statue of King William III had been attacked and defaced on numerous occasions throughout the course of the nineteenth and early twentieth centuries, but was dealt the ultimate blow in 1928. During the annual Remembrance Day commemoration on 11 November, the statue was badly damaged in an explosion, and an attempt was also made to destroy the George II monument in St Stephen's Green.[3] While *The Irish Times* stated that 'the efforts appear to have been made simultaneously, and without any great success, for although some masonry of the pedestals was displaced, the statues themselves remain proudly erect',[4] the damage was in fact much more acute. A later report commented that:

> a large gang of Corporation workmen are busy at work erecting props around King William's statue. It appears that, following the explosion of Sunday morning, there is some slight risk, owing to the damage done to the pedestal that His Majesty and his charger may topple onto the street. The authorities are taking no chances, and the monument is to be firmly held in position by the props until such time as repairs can be made.[5]

On 30 November 1928 the statue was removed from College Green under orders of Dublin Corporation, and rather than being returned to its plinth on College Green it remained in a Dublin Corporation lumber depot.[6] The site remained empty until 1945 when it was reserved for an intended statue to Thomas Davis.[7]

In the same year that the statue of William III was bombed, the figure of George I that once stood once in a prime position on Essex Bridge, one of the city's key bridges, but had been demoted to a garden at the rear of the Mansion House, was sold. As the Irish correspondent for *The Sunday Times* reported on viewing it through a key-hole in the gate of the Mansion House yard, it stood 'facing the entrance as if ready to move out when the gate was opened'.[8] The fate of the statue was taken up by Thomas Bodkin who had left Ireland in 1934 to become Director of the Barber Institute of Fine Art at the University of Birmingham.[9] He, together with an architect of the Barber Institute, Robert Atkinson, had been searching for an appropriate monument to erect outside the building. Bodkin recognised the merits of Van Nost's statue and seeing it as a work of art rather than a symbol of imperialism he set about securing it for the Institute.[10] In 1937 Bodkin contacted the former president, W. T. Cosgrave, whom he sought to intervene on his behalf. Cosgrave contacted the city manager informing him of Bodkin's plan, and the latter sought the advice of the law agent who saw no legal difficulty in selling the statue.[11] The Lord Mayor, Alfie Byrne, was also consulted and expressed his unwillingness to sell the statue for cash arguing that such a transaction might be 'intrinsically objectionable'. Instead he proposed that an offer be made to the Corporation to exchange the statue for a 'suitable picture or pictures of approximately equivalent value for the purposes of exhibition in the Municipal Gallery'.[12] The Lord Mayor wrote to Bodkin and suggested that the best approach would be to write to Dublin Corporation informing them:

> that you have a nice piece of art, which is not in possession of our Corporation, and which you could present . . . in return for the statue of George II which is now at the rear of the Mansion House.[13]

Bodkin proposed the artistic exchange or alternatively the lodgement of a sum representing the value of the statue to be expended in the purchase of a work or works of art for the Municipal Gallery which was without funds for the purchase of modern pictures for the city.[14] When the matter came up for consideration by the general purposes committee of the Corporation in September 1937, its members agreed to sell the statue.[15] A sum of £500 was agreed between the parties as an appropriate price and the monument was removed to the Barber Institute in Birmingham where it now stands (plate 7.1).[16]

The statue of George II in the centre of St Stephen's Green met with a similar fate to that of William III. On 13 May 1937, the day after the coronation of King George VI, the monument was destroyed in an explosion (plate 7.2).[17]

> Early yesterday morning the statue of King George the second in St. Stephen's Green Dublin was blown up by an explosive surreptitiously placed in position during the night. . . . Shortly after eight o'clock . . . a deafening explosion shattered the quiet of St Stephen's Green, wrecking many windows in the surrounding houses, and causing a good deal of alarm among residents and passers-by. The bronze equestrian statue of King George the Second which had stood in the centre of the Green since 1758, was blown to pieces, and fragments of the granite were hurled thirty yards away.[18]

Plate 7.1 **The statue of King George I**
Outside the Barber Institute, Birmingham

While the site was later considered for a statue to Thomas Davis, this was eventually erected on College Green in 1966. The 'war' against statues of empire continued apace, however. The Gough statue in the Phoenix Park was mutilated on Christmas Eve 1944, when, as reported in *The Irish Times*, 'The head and sword were sawn off. Passers-by on Sunday morning noticed that the head and sword had been removed. They have not yet been recovered.'[19] The statue was badly damaged in an explosion in 1957 and was subsequently sold.

From Dublin to Sydney: the fate of the Queen Victoria Monument

We have certain symbols to which people object, such as oaths and privy councils, and other things, and I think that the most prominent symbol, which it would be most advisable to remove, is the one which is in the power of the parliamentary secretary to remove. In this year, particularly, it is very unsuitable that this oriental potentate, First empress of India should be displayed in front of our legislative assembly.[20]

Plate 7.2 **The destroyed statue of King George II, St Stephen's Green**
The Irish Times, 14 May 1937

Probably one of the most famous of the statues to be removed in the years after independence was that dedicated to Queen Victoria. Removing the monument from outside Leinster House was urged repeatedly after 1922.[21] In August 1929, *The Star* published an article calling for the removal of 'An ugly monument'.[22]

> The monument . . . is a particularly ugly piece of statuary. Architecturally it is out of keeping with its surroundings, and its monstrous massiveness obscures the view of Leinster House from Kildare Street. Visitors, Irish and Foreign, to the beautiful building in which the Oireachtas meets, are unanimous in condemnation of the hideous pile . . . The monument is, in addition, . . . repugnant to national feelings. It was erected, not so much 'to perpetuate the memory of a great British sovereign' . . . as to misrepresent the national outlook towards British rule in Ireland in the eyes of strangers. Surely it is not claimed that the representatives of the Irish people would be exceeding their rights in having the statue removed? . . . monuments like those to Victoria and Nelson in the streets of the Capital evoke memories which it would be in the best interests of all to forget.[23]

Some days later a report in *The Irish Times* suggested that the statue was to be removed from Oireachtas buildings:

> on the grounds that its continued presence is repugnant to national feeling, and that, from an artistic point of view, it disfigures the architectural beauty of the parliamentary (Oireachtas) buildings.[24]

While the monument was subject to much discussion in the newspapers, it eventually reached the Dáil when the issue was raised by Tomás Ó Maoláin to the Minister for Finance, Ernest Blythe. He asked, 'whether it is intended to remove the Queen Victoria statue from the front of Leinster House'. Mr Blythe replied, 'The statue in question is not regarded as a valuable or attractive work of art; nevertheless, it is not thought that its effect on popular taste is so debasing as to necessitate the expenditure of public funds on its removal.'[25] The voices calling for the removal of the monument gathered volume and in a lengthy article in the *Irish Press* Norbert Johnson observed:

> nothing recommends the retention of that statue. There are not three members of the Dáil whom it does not offend in their national or democratic instincts. There is not one member of the Dáil whose artistic sense it does not outrage. There are at least eighty-five members of the Dáil to whom that presence out- side the legislative chamber of Irish nationalism is an affront to their conscience and to the people whose ideals they are elected to cherish. Take it away! Ye who have so sorely won the power against the forces it commemorates.[26]

In the Spring of 1937 the issue once again came before the Dáil[27] while *The Irish Builder* remarked:

> it is perhaps not too much to hope that some day a fountain will replace the present statue of Queen Victoria in the courtyard of Leinster House. As a work of art the statue leaves much to be desired and its size and detail are entirely out of keeping with the west façade of Leinster House, a good example of the Georgian architecture for which the city is so widely noted. We think that there was a proposal at one time, to erect a fountain with a low surround in this courtyard which would at once open to view the main entrance to the building and, if carefully designed, add considerably to the general effect of the courtyard. At the same time the gateway from Kildare Street could advantageously be considered with the object of removing the two flanking pavilions, which also serve to conceal so much of a pleasing elevation.[28]

A year later in 1938 the *Dublin Evening Herald* reported that 'Vancouver wants to erect a statue to Queen Victoria, and arrangements are being made to start a fund for the purpose . . . we might save Vancouver a good deal of unnecessary expense . . . by presenting to that city the statue of Queen Victoria which is outside Leinster House.'[29]

Dublin Corporation also added its voice to the calls for the removal of the monument. In 1943 it passed a resolution unanimously calling for the removal of the statue of Queen Victoria from Leinster House:

> It was moved by Councillor O'Maoláin and seconded by Councillor Mrs. Clarke 'That having regard to the changed political status of this country we, the elected representatives of the capital city of Ireland request the government to have the monument of the late English Queen Victoria removed from its site in front of the Parliament House and to have a fitting memorial to Lord Edward Fitzgerald erected in its stead in front of his family home.'[30]

While county councils from around the country offered support to the calls by Dublin Corporation for the removal of the figure, with some suggesting that it be removed to Belfast where it could occupy a space in the grounds of Stormont, the issue was eventually resolved in June 1948 when a scheme was approved 'for the provision of parking accommodation for motor-cars at Leinster House'.[31] On 22 July 1948 the statue was finally removed to the grounds of the Royal Hospital in Kilmainham, a fitting gesture as many saw it, to mark the year in which Ireland left the Commonwealth to become a republic. *The Irish Times* cover photographfor the day shows the statue being winched from its pedestal by a crane and removed to Kilmainham. Ministers, Deputies and others watched as the statue of the Queen was 'removed on her back as the figure was too tall to pass through the gates of Leinster House' (plate 7.3).[32]

Following the removal of the statue a number of requests were received by the OPW from foreign interests who wished to purchase it.[33] Interest from Canada was particularly strong, with requests from the city councils of London, Ontario and Victoria, British Columbia.[34] Neither, however, could raise the funds necessary to ship the statue to Canada and the file on the matter was subsequently closed.[35]

In 1983, a worldwide search began in Sydney for a suitable statue to be erected in a position adjacent to the Queen Victoria building (QVB) in the Australian city. The QVB had been built in 1898 as a monument to the British monarch. Having investigated the options in a number of countries, Neil

Plate 7.3 **The removal of the Queen Victoria statue**
OPW files: Queen Victoria monument

Glasser, the Director of Promotions at the QVB, travelled to Ireland to view the statue of Queen Victoria, then located in the museum overflow store at Daingean, County Offaly. He found the statue:

> behind a brick wall belonging to a derelict reformatory school seated on damp ground, exposed to the inclement weather of Ireland. For forty years she had served as a favourite perch and nesting place for the local birds. Bush and brambles had sprung up all around and although undamaged, her time outside had left her discoloured with the bronze well hidden under a coat of black and greens.[36]

Glasser wrote in a letter to Garret FitzGerald, the then Taoiseach, that:

> I have just today returned from viewing [the statue of Queen Victoria]. I am overwhelmed and excited by its sheer majesty and magnitude. It is in my opinion perfect for the city of Sydney's restored Queen Victoria building, which will be opened later in the year . . . The statue would be in pride of place in the main area in front of the QVB, directly facing our town hall. I would be recommending the area be named the Irish enclave.[37]

The transfer of the statue did not proceed without some opposition, however. The Minister for Finance argued that it should be placed in a museum in Ireland instead. A memo from his secretary to the OPW made clear the views of one sector of the administration:

> It may be fashionable at present times to regard late nineteenth and early twentieth century Anglo-Irish artefacts as skeletons in the National cupboard. My suggestion would be to leave the cupboard closed rather than toss the contents out to anyone who happens to ask for them.[38]

Later, in another memo, the private secretary of the Minister for Finance to the Minister responsible echoed these sentiments:

> My own view is that this statue should be displayed in Ireland perhaps indoors in a Museum. The Queen did receive a warm welcome in Dublin and Dan O'Connell proposed many a loyal toast to her. In the context of the Anglo-Irish agreement, 'two traditions' and so forth, I see danger in publicly jettisoning a symbol of the second tradition. If the matter goes to cabinet the advice should be that we do not dispose of it.[39]

In September 1986, despite these objections, the government decided to give the monument to the people of Sydney, on the occasion of the bicentenary celebrations.[40] The statue was subsequently removed to Sydney (plate 7.4) and it now stands in the Bicentennial Plaza, facing Sydney's Town Hall, with the inscription:

> At the request of the city of Sydney
> this statue of Queen Victoria
> was presented by the people of Ireland
> in a spirit of good will and friendship
>
> Until 1949 it stood in front of
> Leinster House, Dublin, the seat of the Irish parliament
> Sculptured by John Hughes, R.H.S.
> Dublin 1865–1941
> Unveiled on the 20th December 1987 by
> Sir Eric Neal, Chief Commissioner.[41]

The fate of Nelson's Pillar after 1922

Ever since its unveiling in 1809, Nelson's Pillar had exercised the minds of many, evoking an uneasy combination of aesthetic admiration and political disquiet. Following Independence, calls mounted for the removal of the column in the interests of the civic improvement of the O'Connell Street area and groups such as the Dublin Citizens' Association, the Dublin Tenants Assocation and Dublin Corporation threw their weight behind the proposal. It was argued that 'the Pillar is most unsuitably placed, a great obstruction to traffic, and forms an objectionable barrier, severing the north from the south side of the city, with very ill results for the trade and commerce and the residential amenities of Dublin'.[42] In the Senate, W. B. Yeats suggested that:

> if another suitable site can be found Nelson's Pillar should not be broken up. It represents the feeling of Protestant Ireland for a man who helped to break the power of Napoleon. The life and work of the people who erected it is a part of our tradition. I think we should accept the whole past of this nation and not pick and choose. However it is not a beautiful object.[43]

A more novel solution was proposed in a letter to the *Irish Builder* when it was suggested that:

Plate 7.4 **Queen Victoria monument, Sydney**

every statue in the Dublin streets should be taken down and re-erected in Merrion Square, to be hereafter known as 'Monument Park'. . . The Pillar could be erected in the centre, with Nelson overlooking the sea and the vast British Empire, and keeping a blind eye on the doings of Leinster House. O'Connell and Smith O'Brien by their proximity might improve the tone of our modern legislators, and Parnell remind them that there are no bounds to the march of a nation.[44]

Elsewhere it was argued that:

the pillar should not be removed until ample provision had been made for its re-erection on a suitable open site. To merely pull down the pillar without provision for re-erection and leaving all the worse monuments would be a retrograde movement, and we hope it will not be carried into effect.[45]

Consequently, a range of more suitable locations were proposed, among them the Phoenix Park, the centre of Merrion Square and the Hill of Howth, where, it was claimed, the monument would serve 'as an inspiration to all future lovers of the Empire. Liverpool has already made a claim for the statue and column, and the probabilities are that the highest bidder will get them'.[46]

The underlying political significance of removing Nelson's Pillar came to the fore during a Dublin Corporation debate in 1931 when some councillors pointed to the shame inherent in having 'Nelson in the middle of the capital city, while such Irishmen as Red Hugh O'Neill, Patrick Sarsfield, Brian Boru, and Wolfe Tone had no memorials. The deeds of such heroes should not be concealed from the youth of Ireland!'[47] Some years later in 1949 the authority passed a resolution 'that the statue of Lord Nelson in O'Connell Street should be removed and replaced by that of Patrick Pearse'.[48] At the same meeting, however, the Law Agent pointed out that although a public monument, Nelson's Pillar 'is private property, and that if the Corporation desire to acquire it, it can only do so by agreement with the Trustees, or under powers conferred by a Special Act of the Oireachtas'.[49] The Trustees remained unwilling to relinquish their power and consequently speculation over the future of Nelson's Pillar persisted. In 1954 a number of letters were addressed to Dublin Corporation from various organisations. The Dublin Brigade of the Old IRA, for example, proposed that, 'The Dublin Corporation seek legislation for the removal of the Nelson Pillar.'[50] The Port St Anne Society in County Down suggested that:

If it is ever decided to take down the above monument, then in that event, the Port St. Anne Society would be prepared to negotiate for the complete figure of Nelson on top of the Pillar. This figure would be an ideal one for us to have on behalf of our work for the restoration of Killough, Port St. Anne harbour, and each year we would arrange a festive meeting at which the Dublin Corporation and its people would be toasted by a good supply of the best from Guinness's Brewery.[51]

A year later the 'Australian League for an Undivided Ireland' added its voice to the debate when a letter was submitted to the Corporation outlining the terms of a resolution adopted by its Executive:

That the Lord Mayor and members of the Corporation of the City of Dublin be requested to give consideration to the removal of the statue of Lord Nelson from the pillar in O'Connell Street and to its replacement by a statue of Theobald Wolfe Tone, originator of the republican movement in Ireland. It would be appreciated if this matter were given consideration, the promoters of the cause for a United Ireland in these far off parts await the result of your deliberations.[52]

The Arts Council of Ireland also wrote to Dublin Corporation suggesting that any changes to the monument should be confined to changing the statue only.[53] In 1955 the Corporation eventually proposed that 'The Dublin city council request the Trustees of Nelson's Pillar to grant permission to remove the statue of Lord Nelson and that the said statue be placed in the National Museum or other place named by the Trustees.'[54] A letter was subsequently received, however, from one of the Trustees of the monument refusing permission to remove the statue, stating that:

> The Trustees find themselves debarred from granting the Council's request by the terms of their trusteeship of this Monument which impose upon them the duty to embellish and uphold the Monument in perpetuation of the object for which it was subscribed and erected by the citizens of Dublin. They consider that they have no power to vary or depart from the duty thus laid upon them. In these circumstances it is not possible to accede to the City Council's request.[55]

At Government level the question of removing the monument also arose on a number of occasions. During his second term of office the Taoiseach, John A. Costello, argued that on historical and artistic grounds the monument should be left alone. He invited Dr Thomas Bodkin, former director of the National Gallery and the Director of the Barber Institute of Fine Arts, Birmingham, to give a lecture on the pillar. Bodkin described its architectural merits and instanced cities that retained monuments to individuals who had fallen from favour, including a monument dedicated to Tsar Peter the Great in Communist Leningrad. He was scathing in his criticism of the suggestion that Nelson be replaced by a statue of the Blessed Virgin: 'I can't help thinking that she would not like to take charge of a column that was subscribed for and erected to the memory of someone else.'[56] When Sean Lemass became Taoiseach he suggested that an appropriate replacement for the figure of Nelson would be that of St Patrick, coinciding with the Patrician Year of 1961.

Despite the various official attempts to have Nelson removed, the fact remained that the trustees of the Pillar were legally bound to safeguard its position. The fate of the monument was eventually taken out of their hands, however, when it was badly damaged in an explosion one month before the 1916 Golden Jubilee celebrations (plate 7.5):

> The top of Nelson Pillar, in O'Connell Street, Dublin, was blown off by a tremendous explosion at 1.32 o'clock this morning and the Nelson statue and tons of rubble poured down into the roadway. By a miracle, nobody was

Plate 7.5 **Nelson's Pillar after the bomb in 1966**

From *The Irish Times*, 9 March 1966

injured, though there were a number of people in the area at the time . . . Gardai set up a cordon around the city and checked on the movements of members of the Republican movement, but it appeared that the pillar had been shattered by an explosion which had been set some time previously. . . The demolition of the pillar was obviously the work of some explosives expert. The column was cut through clearly just below the plinth, and the debris fell closely around the base of the monument, with some stone being hurled just as far as the entrances of Henry Street and North Earl Street.[57]

The editorial of *The Irish Times* on 9 March 1966 observed that:

Ever since it went up it has been the subject of controversy. It blocked the traffic; interrupted the view of the street, and when we attained our sovereign independence, it was as odd to have Nelson up there as it would be to have had an Austrian general's statue in the principal thoroughfare of Milan.[58]

Shortly after the initial explosion, which was widely thought to have been carried out by a splinter group of the Republican movement known as Saor Uladh, the Government, under the direction of Sean Lemass, gave authorisation for the blowing up of the remainder of the pillar by the Army, occasioning in the process considerably more damage to the surrounding area than the initial blast. After the demolition, the Nelson Pillar Bill came into being, providing legislation that dealt with the inevitable legal aftermath of the demolition. By the terms of the Nelson Pillar Act the trustees were awarded £21,750 in compensation in for the Pillar's destruction and its removal, while other compensation covered the loss of admission revenue and legal costs. The site on which Nelson once stood was vested in Dublin Corporation and the Nelson Pillar Trust was accordingly declared terminated. Section Four of the Nelson Pillar Act of 1969 conferred on the Pillar's trustees an indemnity to cover any suit that might be brought against them for not restoring the monument.

The debates that accompanied the reading of the Nelson Pillar Bill in both the Dáil and Seanad make for entertaining reading. During the second stage reading in the Seanad the debate was virtually monopolised by Owen Sheehy-Skeffington. He touched on a feeling shared by many Dubliners in his statement that:

When in 1966 the pillar was half blown down by a person or persons unknown, I, as a Dubliner, felt a sense of loss, not because of Nelson – one could hardly see Nelson at the top – but because this pillar symbolised for many Dubliners the centre of the city. It had a certain rugged, elegant grace about it . . . The man who destroyed the pillar made Dublin look more like Birmingham and less like an ancient city on the River Liffey, because the presence of the pillar gave Dublin an internationally known appearance.[59]

The contentious debate about the fate of Nelson's Pillar in post-Independence Ireland offers a number of insights into the powerful role of public statuary in cultivating narratives of identity. The monument acted as a focus for the divergent views of Dubliners through a period of radical political and social change. While for some the Pillar had become a jarring symbol of colonial rule, for others it constituted an obstruction to the flow of traffic through an ever-expanding city. With the passage of time it became a popular meeting-place and viewing-point, the terminus of the tramway system and a symbol of the city centre that effectively transcended any political connotations. As one commentator put it shortly after the destruction of the monument, 'There may have been different views on his presence, but to us he was a land-mark. Once we spotted Nelson we knew where we were.'[60] This perhaps explains why the State, despite the intermittent calls that were mounted for its removal on either political or traffic grounds, never formally sanctioned such a course of action. Instead a dissident group seized upon its political symbolism and in an iconoclastic gesture fundamentally altered the iconography of O'Connell Street for ever.

The destruction of the Gough, Carlisle and Eglinton and Winton monuments

Dublin statues are being gradually weeded out. In July 1948, the statue of Queen Victoria which had stood outside Leinster House was removed and is now in the Royal Hospital Kilmainham. The following year a statue of King William III had to be removed from College Green by Dublin Corporation and is stored somewhere in the Mansion House grounds. Lord Gough after three attempts was eventually blown up in the Phoenix Park in July last year; and the Carlisle memorial in the Phoenix Park suffered a similar fate in July this year, and a 1914–18 war memorial in Limerick was blown up in August last year.[61]

Just as had been the case in the decades prior to the establishment of the Irish Republic, the desire to eradicate the colonial connection by destroying or removing the monuments dedicated to figures or events associated with the British Empire continued in the 1950s. Invariably the statues were removed by force of bomb rather than governmental intervention, and the vulnerable position in which they stood was underlined during a Dáil debate in 1961, in which the extent of the damage inflicted upon some of the statues redolent of the connection with the British Empire was made clear (table 7.1).

While the statue of Queen Victoria was removed under the direction of the Irish government in 1948, a gesture which symbolised the political transition that was to take place a year later, other monuments that honoured figures drawn from the realm of the British monarchy and military were removed by force of bomb. Among these stood the statue of Viscount Gough in the Phoenix Park which had been erected in 1880. The equestrian statue had been subject to attack on a number of occasions.[62] In October 1956 paint was poured over it and a month later it was damaged in an explosion (plate 7.6). A more severe explosion in July 1957 shattered the monument. *The Irish Times* reported that:

> the massive Gough equestrian statue in the Phoenix Park, Dublin, was hurled to the ground early this morning by a powerful explosion and shattered. The explosion was heard over a wide area . . . within minutes a large crowd gathered at the scene . . . they found the horse . . . lying on its side ten feet from the base of the memorial. On the memorial steps was Viscount Gough's head.[64]

Almost exactly a year later on 28 July 1958, the statue dedicated to the Earl of Carlisle, also located in the Phoenix Park, was:

> blown from its pedestal and damaged badly . . . It was only about 100 yards from the Civic Guard headquarters, which are situated near the North Circular Road entrance to the Park, and within about half a mile of both McKee Military Barracks and Collins Military barracks . . . The statue is believed to have been blown high into the air, judging from the fact that it was embedded two feet in the soil beside the pedestal. The base and one side of the face were broken.[65]

Once again no individual was prosecuted for the destruction of the statue although suspicions were directed at the republican movement (see plate 7.7). This was despite the statement that was issued by the Irish Republican Publicity Bureau shortly after the explosion. 'We have been asked to state that the Irish Republican Movement was not responsible for the Carlisle Monument explosion in the Phoenix Park Dublin, early today, and no member of the

Plate 7.6 **Gardaí patrol O' Connell Street in the aftermath of the explosion at Nelson's Pillar**
Wiltshire Collection, NLI

Movement was involved in the incident.'[66] Under the heading 'Petty Vengeance', *The Irish Times* commented:

> It is useless to look for logic in these irresponsible actions. The claim of high-mindedness and sincerity which sometimes is made for those who continue to tilt the British wind-mill withers in the face of outrages such as yesterday's. As a result, the problem ceases to be one of inquiry into motives and becomes one of prevention at almost any cost. Some of the memorials which have been the object of recent attacks may be aesthetically valueless; nevertheless they all form an integral part of our history. To hold otherwise would be to grossly distort the historical relationship of the neighbouring islands. In the eyes of the world, every such act of vengeful destruction can serve only to confirm the false view that we are an irresponsible nation; that, in the popular phrase, 'the country isn't half settled'.[67]

A month later the monument dedicated to the Earl of Eglinton and Winton in St Stephen's Green was also shattered in an explosion. It had been

Table 7.1 **Damage to public monuments, 1952–62**

1952–3   Wellington Monument, Phoenix Park
Hand and sword broken off plaque. Replaced at a cost of £5. The same hand and sword were stolen three months later and were not recovered.

1953–4   Wellington Monument, Phoenix Park
Part of Lightning Conductor and portion of bronze moulding removed. Estimated cost of replacement approximately £130.

1954–5   Eglinton Statue, St Stephen's Green
Defaced with paint. Cleaned at a cost of £17.

1955–6   Wellington Monument, Phoenix Park
Part of lightning conductor removed. Repairs cost £9 10s.

1956–7   War Memorial, Islandbridge
Base of memorial cross damaged by explosion. Repaired at a cost of £30.

1957–8   Gough Statue, Phoenix Park
Horse and rider blown from pedestal. Not replaced.
Estimate of cost of repair not made.

1958–9   Carlisle Statue, Phoenix Park, Eglinton Statue, St Stephen's Green
Both statues were destroyed by explosion. Not replaced.
Estimates of cost of replacement not made.

War Memorial, Islandbridge
Base of memorial cross damaged by explosion. Repaired at a cost of £5.

Phoenix Monument, Phoenix Park
Defaced with paint. Cleaned at a cost of £9 10s.

1959–60 Kettle memorial, St Stephen's Green
Bust removed from pedestal. Repaired at a cost of £10.

Wellington Monument, Phoenix Park
Defaced with paint. Cleaned at negligible cost.

1961–2   Mangan, Kettle and Markievicz, St Stephen's Green.
Defaced with distemper. Cleaned at negligible cost

From *Dáil Debates*, 1961, cols 1324–5, no. 19

Plate 7.7 **The Gough statue in the Phoenix Park after the bomb in 1957**
From *Dublin Evening Mail*, 23 July 1957

The Carlisle monument which was blown off its pedestal in Phœnix Park, Dublin, by an explosion early yesterday morning.

Plate 7.8 **The Carlisle statue in the Phoenix Park after the bomb in 1958**
From *The Irish Times*, 29 July 1958

daubed with green paint a year earlier, but the early morning bomb on 26 August 1958 saw its permanent removal from the sculptural landscape.[68] A statement issued by the Irish Republican Publicity Bureau declared that no member of the movement was involved in the blasting of the statue and went further by stating that 'The creation of such incidents is against Republican policy, which is directed against British occupation of our country.'[69]

Over the course of the ensuing decades a number of requests for what remained of the Gough monument were received by the Office of Public Works, among them one from a distant relative of Gough.[70] In 1984, Robert Guinness of Straffan, County Kildare set about purchasing the statue and sought to re-erect it in the grounds of a large house in Britain. The Minister for Finance, with whom responsibility for the statue rested, declared that he had no objection to the request and the statue, together with the Carlisle monument, were sold to Guinness on 19 June 1984 for £100. It was removed from the grounds of the Royal Hospital Kilmainham two years later in 1986.[71] Controversy arose, however, when the actions of the OPW became public knowledge. Indeed opposition was such that Guinness offered to return the statue on the proviso that it be re-erected in a public position in Ireland, although this never happened.

Conclusion

The 'Irish-Ireland' ethos which successive governments sought to cultivate in the decades following the establishment of the Free State was conveyed not only in the erection of politically appropriate statuary, but also in the removal from the public thoroughfares of the imperial iconography. It should be borne in mind, however, that governments in post-independence Ireland had not developed a policy specifically aimed at the removal of these statues, nor did these governments sanction, tacitly or otherwise, their destruction. Rather, they were removed sporadically by dissident elements.[72] In fact, in 1938 de Valera expressed his concern about the fate of the various imperial relics stated and that it was not government policy to remove sculpture solely because it was associated with the British regime. On the contrary, he argued that 'There may in some cases be reasons of historical or artistic interest which would make it undesirable to take such action.'[73] There were other dissenting voices, among them another Taoiseach, J. A. Costello. He argued during the 1940s that those public monuments redolent of the previous regime should be retained as reminders of the struggle that Ireland had come through. These 'destructive tendencies' demonstrate nonetheless the potent symbolism of the public statue in its removal as much as in its unveiling. In common with countless other countries that have emerged into independent political contexts the removal and or destruction of public monuments delivers a cogent symbolic message. For the very qualities that make public statues so valuable in building popular support for one regimes also make them a target for destruction when that regime falls.

CHAPTER EIGHT

Street naming and nation building in Dublin

Nation building and nomenclature

In cultivating a narrative of national identity after independence, the names ascribed and re-ascribed to the city streets as well as to the newly expanding suburbs proved significant. Although efforts had begun before 1922 to set about changing the names of many of the capital's streets, the achievement of independence gave new impetus to this process. Throughout the decades after 1922 the municipal authority became engaged in a process of renaming the capital's thoroughfares. While initially it set out with a grand plan to rename en masse a swathe of names, this course of action did not prove feasible for a number of reasons, not least the trading difficulties that street name changes engendered and which turned many local residents against them. What is also noteworthy is the choice of names adopted by developers and sanctioned by Dublin Corporation in the expanding suburbs after 1922. The development of suburban housing estates in pockets of land throughout the greater Dublin area provided a fresh canvas upon which a symbolic layer of meaning was to be inscribed.

Dublin Corporation and renaming streets

In 1921 just before the formal creation of the Irish Free State, Dublin Corporation set about orchestrating a concerted policy of street renaming (table 8.1). Members of the assembly sought to remove those names which could be interpreted as signifiers of an earlier age of empire and instead introduce the names of Irish patriots to the capital's street directory, among them Pearse, Connolly and Macken. These motions were later rescinded, however, and it was moved instead:

214

That a committee be appointed by the Council, consisting of gentlemen with expert antiquarian, topographical, and lingual knowledge, to take into consideration the general question of Dublin street nomenclature, and to submit for the consideration of the Municipal Council any recommendation in connection therewith which they may deem desirable.[1]

The deliberations of this committee were published a year later. The report suggested a major programme of street and bridge renaming in the capital, and recommended the incorporation of the Irish language in a more forthright manner than ever before.[2] The nationalist agenda of the authority is revealed both in the streets that were proposed to be renamed and also in those that were to be left in place (table 8.2).

Table 8.1 **Proposed street name changes, 1920**

Old name	Proposed name
Great Brunswick Street	Pearse Street
Brunswick Place	Pearse Place
Queen's Square	Pearse Square
Queen's Terrace	Pearse Terrace
Townsend Street	James Connolly Street
Great Clarence Street	Macken Street
Clarence Place	Macken Place

Compiled using Minutes, 1920, no. 598

Table 8.2 reveals that those names which were deemed fitting for the streets of the independent capital were either ancient in origin, pre-dating the centuries of British rule, or they referred to a landmark building, a nationalist figure or, as in the case of 'Merrion', 'the name being popularly associated not with the individual but with the neighbouring district of Merrion', although it might have been renamed given its association with an eighteenth century Anglo-Irish landowner. Similarly, 'Anglesea', although named after another landowner, had become 'associated with the island of Anglesea'.[3] Although these streets were not to be renamed, it was recommended that the Irish version of the name be given precedence over its English counterpart, a directive which was also to apply to those names which were changed, as outlined on tables 8.3 and 8.4 below.

Table 8.2 **Street names to be left unchanged, 1921**

Abbey Street
Anglesea Street
Aughrim Street
Baggot Street
Christchurch Place
College Green
Dame Street
Dawson Street
Duke Street
Kildare
Lord Edward Street
Mary Street
Merrion Square
Merrion St
Parliament Street
Parnell Street
Trinity Street
Butt Bridge
O'Connell Bridge
Grattan Bridge

Compiled using RPDCD, 1921, vol. I, no. 71.

Had the proposals of the special committee been implemented they would have signalled the dawn of a new era and would have represented a concerted attempt to wipe away the imprint of the previous regime. This was also evident in another aspect of the proposals, which specified that those streets with religious names were to be clearly prefixed with the word 'Saint', namely, Anne Street, James Street, Andrew Street and Stephen's Green. Other streets were to be extended to incorporate a number of pre-existing streets into one, thereby subsuming more than one of the 'colonial' names, as indicated in table 8.5.

The report provoked much interest both within the council chamber and beyond. One letter addressed to the council stated that Dublin had inherited a legacy of street names from a period in the eighteenth century when the great residential streets on the North side – Sackville Street, Gardiner Street, Gloucester Street, Mountjoy Street and Square, Rutland Square – were constructed and which 'perpetuated the memory of an alien aristocracy and ascendancy class':

A general examination of the place names shows that no less than 19 are derived from Kings and Queens and their families, 49 from Viceroys and their families, while probably upwards of 120 may be attributed to State Officials, Lord Mayors, noblemen, property owners, and their connections. Many of these names represent individuals, who so far as they had any connection with Ireland or its capital, may be said to have represented English influence, were hostile to the national spirit and development of the people, and were identified with the repressive government of the period. This system of street nomenclature was a determined policy pursued by the upholders of foreign domination to uproot Irish national tradition and to Anglicise the outlook and habits of the Irish people. We doubt whether any other capital city of a country similarly circumstanced submitted so long to such denationalising influences. At least we think there are few cities in the civilised world where the names of tyrannical governors, oppressors, and despoilers have been honoured to the same extent as in our own city of Dublin.[4]

A memorandum submitted by Charles McNeill, which was appended to the report, points by way of contrast to what he saw as the unsuccessful propagandist value of street names. He observed that the practice of street naming in Dublin had been introduced 'from a foreign country after the Stuart restoration, but hardly came into use until after the year 1690, when it was systematically used to proclaim the triumph of the revolutionary party . . . Names given in the spirit of the party are . . . lifeless when the spirit has departed.'[5] His view was that street names:

are unsuccessful in the line of propaganda, street names as symbols or vehicles of personal respect are no more suitable than garments such as Spencers, Mackintoshes, Garabaldia, or Wellington boots. They are too utilitarian and familiar, and if there is anything recondite in them it is blurred by constant use . . . Honour by street names recalls the frugal mind . . . it is cheap and may become nasty, as has already been exemplified within our own experience. The virtue of the charitable Mary Mercer was not made more conspicuous by giving her name to what was formerly known as 'Love Lane'; nor has the fame of John Henry Foley been enhanced by imposing his on what was notorious as Montgomery Street; Tyrone gains nothing by taking the place of Mecklenburgh, nor the Corporation by stepping into the shoes of a ci-devant Mabbot. Based on no ground of solid reason the practice is as futile as it is foreign.[6]

In response to McNeill's memorandum, another member of the street nomenclature committee offered a contrasting point of view. He argued that:

> it is only within the last hundred years that the native Irish have begun to exercise any control of what ought to be the national capital. It seems, therefore, to me that, in considering the street names, our main duty is to erase as far as possible the foreign ascendancy names and find suitable native substitutes.[7]

Further memorials regarding the report were also received from both members of the general public and the Oliver Plunkett Association. The former advocated a cautious approach to street renaming, while the latter argued strongly against the renaming of Fitzwilliam Square in honour of Oliver Plunkett, arguing instead that his name should become attached to 'a prominent Dublin thoroughfare adjacent to Dublin castle in whose dungeons the martyr underwent the first imprisonment'. It was agreed that the report be referred on to a committee of the whole house. Rather than adopting wholesale the changes suggested, the authority ruled instead that they be implemented in a number of stages:

> so as to prevent inconvenience to the public and to property owners and householders. If the Council agree that only a limited number of alterations should take place annually, it is desirable that they should make a selection of about half a dozen streets to inaugurate the proposed revolution in street nomenclature.[8]

The report was later carried, subject to the consent of the residents and with an amendment 'That the words "subject to the deletion of those relating to Earl Street, Talbot Street, and Grafton Street" . . . be deleted.'[9] The special street naming committee subsequently produced another report in which they reiterated, despite some opposition, their view that the Irish form of the name should take precedence over the English rendering which was to be placed in brackets after the Irish.[10] The report also suggested a select list of streets for renaming (see table 8.6).

Table 8.3 **Proposals for new street names, 1921**

Old name	Proposed name[1]
Adelaide Road	Incorporated into South Circular Road
Amiens Street	Bohernatra (Strand Road)
Aungier Street	Incorporated into George's Street
Beresford Place	Connolly Place

Old name	Proposed name[1]
Blessington Street	Comeen Street
Bolton Street	Incorporated into Capel Street
Brunswick Street, Great	Pearse Street
Brunswick Street, North	Channel Row
Camden Street Lower and Upper	Incorporated into George's Street
Capel Street	Silken Thomas Street
Cavendish Row	Incorporated into Parnell Square
Clanwilliam Place	Incorporated into South Circular Road
Denmark Street	Incorporated into Gardiner's Row
D'Olier Street	Smith O'Brien Street
Earl Street	Brian Boru Street
Exchequer and Wicklow Street	Wicklow Street
Fitzwilliam Square	Oliver Plunkett Square
Frederick Street	Incorporated into Blessington Street
Gardiner Place, Row and Great Denmark St	Thomas Ashe Street
George's Street	Cahirmore Road
Grafton Street	Grattan Street
Harrington Street	Incorporated into South Circular Road
Henrietta Street	Primate's Hill
Herbert Place	Incorporated into South Circular Road
Henry Street	Incorporated into Mary Street
Leinster Street	Incorporated into Nassau Street
Mountjoy Square	Tom Clarke Square
Nassau Street, Leinster Street and Clare Street	Tubber Patrick Street
Portland Row	Incorporated into North Circular Road
Redmond's Hill	Incorporated into George's Street South
Richmond Road	Tolka Road
Richmond Street	Incorporated into George's Street South
Rutland Square	Parnell Square
Sackville Street	O'Connell Street
Seville Place	Incorporated into North Circular Road
Suffolk Street	Incorporated into Saint Andrew Street
Talbot Street	Incorporated into Earl Street
Warrington Place	Incorporated into South Circular Road
Westmoreland Street	Shanid Street
Wexford Street	Incorporated into George's Street
South William St	Incorporated into Saint Andrew Street
Wilton Terrace	Incorporated into South Circular Road
York Street	Thalia Street

Compiled using RPDCD, 1921, vol. I, no. 71
1   The Irish version of each name change was given precedence in the report

Table 8.4 **Names of bridges for which changes were proposed, 1921**

Old name	Proposed name[1]
Wellington Bridge	Liffey Bridge
Richmond Bridge	O'Donovan Rossa Bridge
Whitworth Bridge	Dublin Bridge
Queen's Bridge	Queen Maeve Bridge
Victoria Bridge	Rory O'Moore Bridge
King's Bridge	Sarsfield Bridge
Sarah Bridge	Island Bridge

Compiled using RPDCD, 1921, vol. I, no. 71
1   The Irish version of each name was given precedence in the report.

Table 8.5 **Proposals to amalgamate street names, 1921**

New name	Derivation
North Circular Road	To apply to the route-way along the canal, including Portland Row and Seville Place
South Circular Road	Incorporated Adelaide Road and Clanwilliam Place
Dorset Street Upper/Lower	To apply to the route-way from Frederick Street to the city boundary at Whitehall. Divided into upper, middle and lower. The upper section to be Upper Drumcondra Road, the middle to extend from Tolka Bridge to the Canal Bridge, and the lower section to consist of Dorset Street, Upper and Lower, to North Frederick Street
George's Street	To incorporate Aungier Street, Wexford Street, Redmond's Hill, Camden Street Lower and Upper and Richmond Street
Capel Street	Incorporated Bolton Street
North Strand	Incorporated Amiens Street

Compiled using RPDCD, 1921, vol. I, no. 7

Table 8.6 **Streets and bridges selected for renaming, 1921**

Street names

Old name	New name
Amiens Street and North Strand	Bohernatra
Great Brunswick Street	Pearse Street
Circular Road, North	To apply to the route-way along the canal, including Portland Row and Seville Place
Circular Road, South	To apply to Harrington Street, Adelaide Road, Wilton Terrace, Herbert Place, Warrington Place and Clanwilliam Place
Earl Street	Brian Boru Street
Grafton Street	Grattan Street
Grattan Street	Grattan Street Little

Bridge names

Old name	New name
Butt	no change
O'Connell	no change
Wellington	Liffey
Grattan	no change
Richmond	O'Donovan Rossa
Whitworth	Dublin
Queen's	Maeve
Victoria	Rory O'Moore
King's	Sarsfield
Sarah	Island

Compiled using RPDCD, 1921, vol. II, no. 209.

Dublin Corporation subsequently voted for the street name changes with the proviso that the residents were also in favour.[11] In only one case, Great Brunswick Street, was there a majority of the ratepayers in number and value in favour of the change.[12] A year later Great Brunswick Street became Pearse Street,[13] while in 1924 it was proposed that Rutland Square be renamed Parnell Square, in honour of C. S. Parnell, after whom Great Britain Street had been renamed in 1911. Consideration of this motion was postponed, however, and it was not until 1933 that the motion was eventually carried.[14] Also in 1924 the motion was passed that, 'the name of Sackville Street be changed to Sráid Ó Conaill

(O'Connell Street) at the request and subject to the consent of the majority in number and value of the rate-payers of the street, as required by section 42 of the Dublin Corporation Act, 1890.'[15] Over the course of the years from 1924 on a number of significant name changes were carried out, notably clustered in the north and east inner city (table 8.7 and figure 8.1), while some streets, which Dublin Corporation sought to alter, remained unchanged (table 8.8).[16]

Dublin Corporation also set about renaming the bridges that linked north and south of the River Liffey in an attempt to articulate the independence of the new state (figure 8.2). The fact that the Corporation could not encounter any residential objections eased the transition. Names were introduced that derived from sources related to Irish nationalist politics, culture or local geographical features.

Street nomenclature: the broader context

One of the major tasks faced by Dublin Corporation in the years after 1922 was the provision of housing in the city suburbs as part of a concerted effort to relieve the slum conditions that engulfed parts of the inner city. The naming of such suburbs became a particular concern of the local authority, the members of which were eager to ensure that suitably appropriate names, derived from Irish antiquity, physical features or associated with religious figures were utilised. Such names served to give a religious and nationalist ambience to public rented accommodation that was often at variance with the tastes of private developers. A range of schemes were developed in 1920s and each employed an interesting choice of nomenclature. In the Marrowbone Lane housing area of the south city, for example, roads were assigned names such as The Virgins Road, Our Lady's Road, Annunciation Road, Morning Start Road, Rosary Road and Ave Maria Road, ensuring that the area became known locally as Maryland.[17] Later in the year the name Loreto Road was substituted for Virgins Road and Lourdes Road for Annunciation Road.[18] In 1934 when the Crumlin housing scheme was nearing completion, roads were assigned names that derived from ecclesiastical sites in Ireland, hence, Clogher, Clonmacnoise, Kilfenora and Monasterboice, among others.[19] In nearby Kimmage, roads were assigned names such as Mount Tallant Avenue, Melvin Road, Neagh Road, Corrib Road, Sheelin Road, Ennel Road, Derravaragh Road and Ramor Road, and in Drimnagh roads were named after Irish mountains and hills, prominent among them Errigal Road, Comeragh Road, Mourne Road, Donard Road, Brandon Road, Slieve Bloom Road and Mangerton Road.[20]

Table 8.7 **Street name changes, 1922–49**

Year	Old name	New name
1922	Amiens Street and North Strand	Bohernatra
	Great Brunswick Street	Pearse Street
	Circular Road, North	Apply through to the canal, Portland Row and Seville Place
	Circular Road, South	To apply to Harrington Street, Adelaide Road, Wilton Terrace, Herbert Place, Warrington Place and Clanwilliam Place
1924	Sackville Street	O'Connell Street
1924	Queen's Square	Pearse Square
	Great Clarence Street	Macken Street
	Wentworth Place	Hogan Place
	Denzille Street	
	Harcourt Place	
	Hamilton Row	Fenian Street
1925	Corporation Street	Store Street
1930	Montgomery Street	Foley Street
1933	Gloucester Street Lower	Sean MacDermott Street
	Parnell Square	Rutland Square
	Findlater's Place and Gloucester Street Upper	Cathal Brugha Street
	Albert Quay	Wolfe Tone Quay
1938	Harold's Cross Bridge	Robert Emmet Bridge.
1938	Dublin Bridge	Father Mathew Bridge
1941	Sarsfield Bridge	Sean Heuston Bridge
1942	Queen Maeve Bridge	Liam Mellowes Bridge
1943	Stafford Street	Wolfe Tone Street

Compiled using Minutes and RPDCD, 1922–49

Table 8.8 **Street name changes proposed, but not carried out, 1924–49**

Year	Old name	Proposed name
1932	Beresford Street	Connolly Square
	Butt Bridge	Congress Bridge
1934	Usher's Island	Heuston Quay
	Island Street	Staines Street
	Mountjoy Square	Thomas Clarke Square
1944	Talbot Street	Sean Tracey Street

Compiled using Minutes and RPDCD, 1922–49

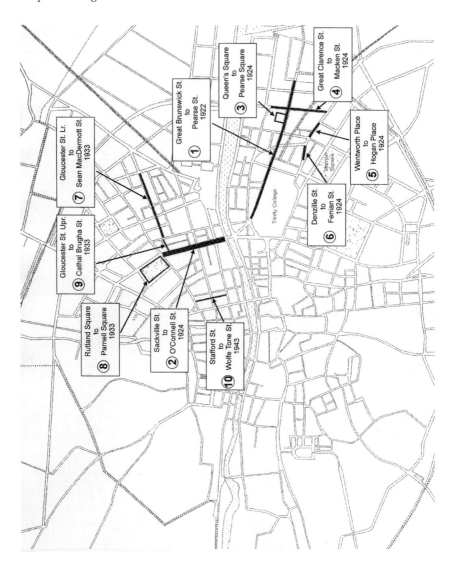

Figure 8.1 **Locations of principal streets renamed, 1922–66**

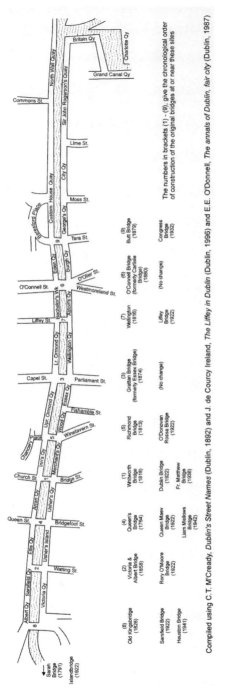

Figure 8.2 **Renaming Dublin's bridges and quays after 1922**

The issues of changing street and town names arose at national government level in the 1940s. During a Dáil debate in 1944 the Minister for Local Government, Sean McEntee was asked by Deputy Larkin:

> whether his attention had been called to the difficulties experienced by nationally minded associations in securing changes of street names with the object of commemorating the memory and works of Irish patriots; and if he will consider introducing legislation to vest in the local authorities . . . the power of making such changes.[21]

While the Minister replied that the matter was under consideration, a year later the issue re-emerged during consideration of 'one of the most extraordinary Bills which ever came before a Parliament'.[22] Section 55 of the bill set out to provide local authorities with the right to change the names of streets at any time. During the second stage reading it became clear that contributors were divided on the issue. Certain TDs expressed their dismay that, while the Minister went 'a certain distance in the right direction', he did not go further. Deputy Allen suggested that:

> There is no reason why, because certain types of people live in a street and probably do not see eye to eye from the national point of view with the Government, or with members of the local authority or with the general mass of the people, they should have the power to prevent a street name being changed; that they should be able to prevent that street being given a good national name in accordance with the view of the mass of the people in a town or district.[23]

Allen went on to argue that the minister should give the local authority the power to change the name of a street where the members of the local authority desired to so do:

> after all, they are the elected representatives of the people. They are in a responsible position and should have that much power without being obliged to get the views of the ratepayers in any particular street . . . I would urge him to go still further and give the local authority, the county council in this case, the power to change the name of any village or town land in their county, if they think it well to do so. I think that the time has arrived when a council should have that power.[24]

In his response to the debate Deputy Liam Cosgrave made no secret of his antipathy to any arrangement that would make possible frequent changes of name.

> While no doubt it is desirable that a majority of ratepayers or citizens residing in a street should, if they desire to change the name of a street, be allowed to do so, I think that this practice of changing the names of thoroughfares can lead to a very ridiculous position.[25]

The Taoiseach went on to observe that:

> In Russia for instance, the changing of names of towns has been a striking feature of the Soviet administration. When an individual happens to be prominent or in power in the country or in a position in which he can change the name of a town, he very often gives it his own name. Then, when he is superseded by somebody else there is another change, the result being that even those who were the most honoured individuals in that country find that the names which they had given to certain places are abolished . . . I think it would be a ridiculous situation if one day a place might be known as Dublin and the next have its name changed to Kerry.[26]

He then questioned the renaming of streets as a form of patriotism, arguing that:

> I think it is a doubtful form of patriotism, because in ten or fifteen years time, as happened in Russia, some other parties may come along and suggest names for these places which they believe are more entitled to be honoured . . . Most of the slum areas of Dublin have the names of very honoured persons attached to them. I do not know whether it ever strikes the people who are anxious to have these name changes carried out that honoured persons names should be attached only to important thoroughfares. There will, of course, be the difficulty that if this desire for change is not limited in some way, we may reach the stage when a person will not know whether he is living in Pat Murphy's Street or Pat McGrath's Street. As a form of patriotism, it is a thing which can be carried to exaggerated lengths and as a form of paying tribute to those who have rendered service to the nation it can assume ridiculous proportions.[27]

The Bill ultimately updated the legislation and stipulated that four sevenths of the ratepayers in the particular street have to give their consent to any name changes.[28] The entire debate demonstrated the strength of feeling that existed, especially among TDs of rural constituencies, for doing away with names that were redolent of the former British administration. The case of Bunclody in County Wexford is a case in point. The town had been renamed Newtownbarry prior to independence and the TD for the area, Denis Allen, expressed his opinion that 'The residents of that village strenuously object to that name . . . If the local authority in Dublin wanted to call this city Baile Átha Cliath instead of Dublin, there is no reason why they should not have the power to do so.'[29]

Conclusion

By 1966 Dublin's symbolic geography differed radically from that which prevailed at the turn of the twentieth century. The country had undergone a transition of enormous political significance which had profound implications for the cultural landscape of the capital. As the colonial legacy was eradicated and Irish nationhood consolidated in the years after 1922, public monuments played a particularly important and highly symbolic role in the cultivation of narratives of national identity. Successive administrations were adept at using monuments, and the choreographed ritual that went with their unveiling, to best advantage. Monuments were erected dedicated to the heroes of the rebellion as part of a strong desire to make tangible the struggle that the country had come through. This struggle was, however, represented as an almost exclusively male one. Women, despite the role that they had played in the independence struggle, are strikingly absent from the monumental landscape and where they are represented it is invariably in allegorical form.

Politics dominated not only the monumental landscape, but also the names that were assigned to the newly laid out streets of the capital's suburbs and in the renaming of streets in the inner city. A series of motions were set in train by Dublin Corporation which sought to rename key city streets in honour of patriots, chief among them the signatories of the 1916 Rising, as well as religious figures or names drawn from geographical features of the Irish rural landscape. A number of the city's main train stations were also renamed after the leaders of the 1916 rising, for example, Kingsbridge became Heuston, Amiens Street station became Connolly, and Westland Row became Pearse Street.

Although there were relatively few new architectural initiatives in the decades after independence, the debate over the shaping of the new capital and the location of key public buildings within it generated much discussion and fostered movements such as the Greater Dublin Reconstruction Movement. The subtleties of these planning initiatives should not be overlooked. Even though successive plans failed in the main to be carried through, they provide an interesting vision of how planners saw the future for the independent capital. The rebuilding of the destroyed public buildings and the debate over the proposed cathedral and Parliament House served to highlight the importance of the architectural medium in the expression of national identity. So too did the building of the Industry and Commerce Building in the 1930s, the first major architectural initiative of the new state, coupled with the 'Shamrock Building' that housed the Irish Pavilion at the World Fair in 1939.[30]

The 1960s proved to be an optimistic decade which built on the success of the state economic development programme from the end of the 1950s. As property developers began to exploit the Georgian heart of the capital, a move not without some symbolic import given that this form of architecture had come to be seen by some as symbolic of the colonial connection with Britain, new office blocks and high-rise flats came to be built. Suburban property development also began to flourish and with that the urban landscape of the capital city changed from being a residential centre with many public buildings, to a commercial centre with some public functions, although increasingly the latter were also moved to suburban locations. The Dublin bourgeoisie continued their move out to the suburbs, driving the need for a new street nomenclature which drew its inspiration not from native political or cultural figures but from the whims of private property developers who tended to draw on generic elements in choosing names.[31]

CONCLUSION

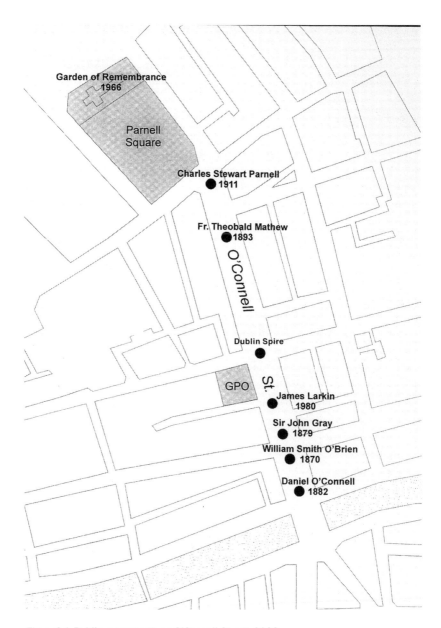

Figure 9.1 Public monuments on O'Connell Street, 2002

CHAPTER 9

The power of space: towards a postmodern iconography

The power of symbolic space

The chapters which make up this work, and the thematic strands of inquiry that they address, confirm that cities are constructed landscapes, shaped by sets of agents that are caught up in a web of economic, social, cultural and political circumstances. The argument that every landscape is 'a synthesis of charisma and context, a text which may be read to reveal the force of dominant ideas and prevailing practices, as well as the idiosyncrasies of a particular author' is particularly true in the context of urban landscapes undergoing the turbulent political transition from colonial dependence to independence.[1] The transformation of the symbolic geography of Dublin before and after 1922 demonstrates that social processes do not occur through time alone but also have an important spatial expression. As Soja argues, 'There are no aspatial social processes.'[2] Our world is only meaningful through the fact that it is spatially as well as temporally extended. It is an accumulation of spaces as much as it is an accumulation of experiences of time.[3]

Such debates about the nature of the connections between place, space, power, language and identity have come to the forefront of geographical research, and the work of Foucault and Lefebvre, both of whom recognised the powerful role played by the spatial alongside the temporal dimension in critical social theory, has been significant. Foucault recognised that 'power requires space, its exercise shapes space, and space shapes social power'.[4] For power to communicate it needs a 'language', and space – together with the elements of which it is composed – constitutes such a language. Lefebvre's *The production of space* echoes this agenda.[5] As the title suggests, his work pursues a theory of space which recognises its dynamic nature as a produced and productive element of society. Using the work of Lefebvre and Foucault as a theoretical base, geographers have set about constructing their own theories of the politics of space, notably Harvey's reading of space within a capitalist framework and Gottdiener's analysis of the urban environment as a socially shared symbol.[6]

In focusing upon particular elements of the urban landscape, this work has demonstrated the power of space in cultivating narratives of identity and giving tangible expression to the past in the present. Public monuments, street names, urban planning and architectural initiatives are fundamentally spatial phenomena, rooted in the domain of the cultural landscape. They each serve in different ways as focal points around which both local and national political and cultural positions are articulated. Statues invariably embody heroic figures of the past or record formative events, making them integral to projects of empire and nation building, as well as focal points of resistance. Street names can be popular political symbols which cultivate national identity and serve as markers of the spatiality of power relationships. Expressions of both dominant and emergent cultures, they often fall victim to a changing political contexts. This work has also established that public buildings and civic design initiatives serve as significant expressions of cultural and political identity. What has been written here, however, is one reading of Dublin's symbolic geography as it had evolved by the late 1960s and as such it is 'a construction that is contingent, partial and unfinished'.[7] It is worth considering in conclusion how the iconography of the cultural landscape began to change in the years following 1966 through the lens of the city's central thoroughfare, O'Connell Street.

From political symbols to public art: the contemporary iconography of O'Connell Street

O'Connell Street has always occupied a position of prime importance and symbolic significance in the life of Dublin and of Ireland; it is after all the main street of the nation, and its 'scale, symmetry, history, architecture and central location bestow a sense of place and civic importance which is embodied in the memory of the people'.[8] After Independence it became an important site of civic ritual and annual spectacle, most especially associated with the commemoration of the 1916 Rising. The monuments erected on the street and in its immediate environs before and after 1922, together with the removal of Nelson's Pillar, created a layer of symbolic significance that was echoed throughout the city. These monumental initiatives played an important role in embodying the dominant ideology of the newly established Free State. In their carefully designed iconography, in the displays of military might and national pride that went with their unveiling, they connected the dead of 1916 with a more ancient and heroic Celtic past that pre-dated the colonisation of Ireland. The cultural landscape came to represent a rich legacy of memories and the shared heritage of a glorious past in order to sustain the narrative of national identity. At the same time, other memories which did not fit so easily with the ideology of the new regime were wiped clean from the landscape in a manner that demonstrates the significance of forgetting in nation building.[9]

Since the 1960s, however, there has been a tangible shift in the symbolic significance of the street, a product of the growing distance from the Independence struggle, the inevitable cultural maturing of the State and perhaps influenced also by the political situation that developed in Northern Ireland.[10] The much more low-key commemoration of the 1916 Rising that occurs today and the absence of any displays of nationalistic military might stand in marked contrast to the earlier State-sponsored 'celebrations' that characterised previous ceremonies.[11] Meanwhile, those monuments that were once erected with such ceremony as symbols of a nationalist ideology would seem to have lost much of their symbolic potency. Many of Dublin's citizens would be hard pressed to name the statues that line O'Connell Street, figures which might now be considered anachronistic features of the cityscape (figure 9.1).

Between the 1970s and the 1990s a number of monumental projects were proposed for O'Connell Street. Some of these made the transition from plan to concrete form, for example, the statue of James Larkin which was unveiled in 1980. Its origins can be traced to 1974 when the Workers Union of Ireland announced their intention to erect a monument to the Trade Union leader James Larkin as part of the fiftieth anniversary of the union. Oisín Kelly was eventually commissioned to sculpt the monument and, although it was planned to unveil it in 1976 to mark the centenary of Larkin's birth, it was not erected until 1980. The figure of Larkin with arms outstretched was unveiled on O'Connell Street Lower, opposite Clery's, the site of the former Imperial Hotel, from where Larkin addressed the assembly of striking workers in 1913. During this time the former Nelson Pillar site also continued to attract much attention with a range of proposals put forward for a possible replacement, among them a suggestion that a statue of Patrick Pearse be erected to commemorate the centenary of Pearse's birth. This aroused lively debate when it was considered by Dublin City Council's planning committee. During the meeting one councillor suggested that the plan proceed because: 'Pearse was the Messiah of Irish independence. He was a cultured man and not a man of violence. This man deserves an honour of some sort.'[12] Although the sculptor Gary Trimble sculpted a model for a 35-foot pillar, topped with the figure of Pearse depicted teaching a group of children, the plan was rejected in 1979 on the grounds that it was out of scale with the streetscape.

The Corporation also rejected another proposal which had been submitted by the Christian Community Centre for a religious statue to serve as a permanent memorial to the then forthcoming visit of Pope John Paul II. On behalf of the 'Concerned members of the Christian Community', Mr T. C. G. O'Mahony suggested the site should be given over to a monument depicting Christ the King, the Queen of Peace, Joseph the Worker and the Holy Spirit, because, he argued:

> the Irish people – at present at a very low economic, social, moral and spiritual ebb – are frantically searching for the answer to pervading dechristianisation and demoralisation, and the visit of this great universal and spiritual leader could mark the advent of a new era of hope for them.[13]

Faced with the prospect of 'innumerable ad hoc ideas from all sorts of pressure groups', the committee adopted a motion tabled by Carmencita Hedderman that a national competition be organised to design a monument to the birth of the Irish nation.[14] Discussions came to an abrupt end in November 1979, however, when Dublin City Council's general purposes committee unanimously decided that the area should be paved over.[15] The decision was made on the recommendation of the Corporation's chief planning officer, Charles Kelly:

> It appeared to us that the development of a simple paved area down the centre of O'Connell Street, aligned with trees and retaining the existing statutory figures but seeking to eliminate all other 'clutter', would result in the development of the street as a really noble thoroughfare.[16]

In 1988 a new monument was unveiled on O'Connell Street, the Smurfit Millennium Fountain or the Anna Livia, designed by Eamon O'Doherty. Sponsored by Michael Smurfit and erected as a memorial to his father, Jefferson Smurfit, just north of the former Pillar site, it heralded the arrival of a new era in the form and content of Dublin's sculptural fabric and more specifically the iconography of O'Connell Street. Made of Wicklow granite and with a female representation of the River Liffey in bronze as its centre point, the monument measured three metres in height and 25 metres in length. It was unveiled on 17 June 1988 to mixed reviews, not least because the public had not been given an opportunity to express any view on the monument and it was exempt from planning control. The Fine Gael TD John Kelly described the water feature as 'a hideous, illuminated fountain fit to recall the era of the Bowl of Light and the "Tomb of the Unknown Gurrier" on O'Connell Bridge which public opinion, after 20 years, succeeded in having removed'.[17] The fountain nevertheless stood as a symbol of the transformation that had taken place in Dublin's monumental landscape in recent years. The controversy that it generated stemmed largely from aesthetic rather than political objections. It was unveiled in a more discreet fashion than its late nineteenth-century pre-decessors on O'Connell Street and occupied only a small number of column inches in newspapers. The erection of the new millennium monument in the centre of Dublin on the location formerly occupied by Nelson's Pillar will serve not only to demonstrate in an even more striking manner the largely apolitical nature of the sculpture which is preferred in the capital for the new millennium, but has also necessitated the removal of the Anna Livia fountain.

From Nelson's Pillar to the 'Spire of Dublin': a monument for a new millennium

After 1966 the site of Nelson's Pillar had been left vacant. In 1987, however, the short-lived Metropolitan Streets Commission suggested that the pillar should be rebuilt. Their advice was not well received, however, and the commission itself was disbanded shortly after. A year later as part of the celebration of the capital's millennium, a competition was launched for the design of a monument to replace Nelson's Pillar. The 'Pillar project' as it became known, was promoted by the Architectural Association of Ireland, the Royal Institute of Architects of Ireland and the Sculptors' Society of Ireland. It brought together fifty architects and sculptors each of whom was requested to make proposals for a new symbol for the city, one which would 'stimulate and contribute to an informed public debate regarding the upgrading of O'Connell Street and establish networks for the future by providing a unique opportunity for architects and artists to work together on the conceptual development of an urban project of significant scale'.[18] The project yielded 17 entries, which were then exhibited in the General Post Office, and the public were invited to vote on their favourite design. The entry that attracted most votes was the 'Millennium Arch' (plate 9.1). Designed by Michael Kinsella and Daniel McCarthy, the triumphal arch was twice the height of the GPO, and incorporated an observation tower and millennium symbol. It was intended to span the centre of the street, and incorporated lifts and spiral staircases in a manner echoing the Arc-de-Triomphe in Paris.

Plate 9.1 **'Millennium Arch', winning design in the Pillar project, 1988**
Designed by Michael Kinsella and Daniel McCarthy.
From J. Graby (ed.), *The monument in the city* (Dublin, 1998)

The 1988 competition, however, was only ever intended to be a generator of ideas and it was never planned to erect the winning design. It was not until 1998 that Dublin Corporation established a competition as part of the redevelopment of O'Connell Street for a new monument. In July 1998 the international design competition was launched with the aim of finding a suitable replacement for the pillar, 'something which would become a new symbol of Dublin for the twenty-first century, just like the Eiffel Tower is to Paris'.[19] In the brief for the competition Dublin Corporation set out the objectives of the competition in design terms, namely, to reinstate a monument which would occupy a pivotal role in the composition of the street and called for a monument with a strong vertical emphasis. A budget of £4 m. was set aside and a clause was inserted that the winning entry would not necessarily be built. The winner was announced in December 1998 (plate 9.2). Picked from 205 entries, the design submitted by the London architect, Ian Ritchie, was for a £3 m. structure of stainless steel which tapers to a light at its pinnacle. The jury argued that the design 'fulfilled the requirements of the competition brief by providing an elegant structure of twenty-first century design. It was also a brave and uncompromising beacon, which reaffirms the status of O'Connell Street'.[20] Ritchie suggested that the 'high and elegant structure' would symbolise 'growth, search, release, thrust – and Ireland's future'. It would be a monument for the new millennium by day and by night, 'a pure symbol of optimism for the future'.[21] It is designed to be 120 metres high, three times the height of the former Nelson's Pillar and twice the height of Dublin's tallest building, Liberty Hall.

The attraction of the monument for some rests in its overtly apolitical nature. It does not appear to have any obvious message political or otherwise. As Pearson has suggested:

Plate 9.2 **'Spire of Dublin' winning design in the millennium monument competition, 1998**
By Ian Ritchie.
Illustration: Dublin Corporation

Perhaps that says something about Ireland and ourselves. Unlike the other monuments around, it's not religious, it's not military, it's not political – which is not to say that it should be any of these things, of course. But it does say something about wealth. I wonder whether in the future people might look back on it and view it as the product of a pointless society.[22]

Nevertheless it is a strongly visual monument, which can be interpreted, as its designer suggests, as a forward looking, contemporary statement, a 'sensational structure which will redefine the city centre and people's perceptions of where that is, quite apart from providing Dublin with a new icon'.[23] On 21 December 2000 the Irish government finally granted approval to Dublin Corporation for the erection of the 'Dublin Spire'. This new monument will shortly be unveiled in O'Connell Street, 'where it will occupy a pivotal role in the composition of the street, fixing the central points of the street beside the General Post Office and closing the vistas from north, south, east and west'.[24]

Amidst this changing political, cultural and social context, O'Connell Street as a totality has retained its status as a central thoroughfare and symbolic heart of the capital. This was recently underscored in the contentious debate surrounding the Dublin Corporation proposal to restrict the right of public protest on O'Connell Street. The proposed by-laws aimed to impose a range of conditions on protests and marches on O'Connell Street, including a 31-day notice period and £2000 deposit for groups over 50. A range of interest groups protested against the motion (which was eventually defeated), arguing that 'O'Connell Street, in which the 1916 uprising's most famous scenes were played out, has epitomised the free and vocal nature of protest in the State, a right enshrined in the Constitution'.[25] The symbolic potency of O'Connell Street was also reiterated in the context of the state funerals that took place in October 2001 for the ten IRA volunteers executed during the War of Independence. In a ceremony that echoed the military displays of a previous era, thousands of people lined the streets of the capital to pay tribute. The processional route took mourners to O'Connell Street where members of the FCA formed a guard of honour and the cortege stopped in front of the GPO. There, a lone piper played a lament before the hearses moved on to the Pro-Cathedral for Requiem Mass. The procession also paused at the Garden of Remembrance, where a minute of silence was observed before the procession proceeded to Glasnevin Cemetery.[26] A week earlier the street had also been a focal point for the commemoration of a more militant form of republicanism when the twentieth anniversary of the 1981 Hunger Strike was marked in the city. Led by a silent row of people who held portraits of the dead hunger-strikers, the procession took participants from the Garden of Remembrance to O'Connell Street carrying on to Leinster House and then back to the GPO where traditional musicians played laments before an address was read to the assembled crowd.[27] Both of these events demonstrate that the symbolic significance of O'Connell Street remains high, even if many of the monuments that line its centre no longer occupy such a central position in the collective memory of Dublin's citizens.

Towards a postmodern iconography

Today, Dublin serves as the capital of what has become known as the 'Celtic Tiger', an over-used and by now clichéd metaphor that signals the economic success of a small independent nation, confident in its Irish, European and global identity. An internationalist, rather than a purely nationalist, spirit has come to characterise the political climate with a number of ramifications for the urban landscape. Throughout the country, urban renewal projects and tax incentives are changing the face of Irish cultural landscapes. Massive out-of-town shopping centres are rapidly out-performing city centre shopping districts. The country, it would seem, has bought into a European and a global iconography that is related to the hegemony of a consumer society. Ireland's 'international vocation' has been promoted and this stands in marked contrast to the isolationist situation that prevailed in the immediate aftermath of independence. This is reflected in the island's urban and rural spaces where new iconographies are currently being inscribed.

All of this would seem to suggest that the cultural context in which Dublin and other mature post-colonial capitals exist has radically changed. Unlike in the decades that followed 1922 when street names, statues and architecture were used almost as badges of identity to exemplify Ireland's status as a Catholic and nationalist nation, today the pluralist nature of Irish identity, contested as it is along a range of new axes of differentiation, has been underlined. This was illustrated in November 1998, for example, on the 80th anniversary commemoration of Armistice Day. Then, a 110-foot high Irish round tower, made from 400 tons of stone from the former Mullingar workhouse, was erected in Belgium. It is located at the Flanders Peace Park in the Belgian village of Messines, where Irish members of the British Army died side by side with their British counterparts in the First World War. Unveiled by the Irish President, Mary McAleese, who was accompanied by Queen Elizabeth II, it marked the first time that the Irish State formally honoured the quarter of a million Irish people who served, and the 50,000 who died, in this war. This event forms a distinct contrast with the events surrounding the erection and fate of the Irish National War Memorial in Dublin's Islandbridge some decades earlier. That commemorative project, which was begun in the 1920s, met with antipathy from many in Free State Ireland who had resented the participation of Irish people in the British war effort. The implacable opposition from some quarters ensured that the project took many years to complete. The memorial park subsequently fell into a state of disrepair and was not restored until the late 1980s when the formal dedication ceremony took place.[28]

Ultimately, the iconography of Dublin's cultural landscape at the dawn of the new millennium is radically different from that of seventy years previously. It nevertheless continues to demonstrate the power of space and reiterates the fact that each generation weaves its world out of image and symbol. Moreover, it provides the geographer with cultural landscapes and symbolic spaces worthy of decoding and ensures that the iconographic method will remain central to cultural inquiry and to the geographer's approach to the description and interpretation of the cultural landscape.

Notes

Abbreviations used in the Notes and Bibliography

AAAG	*Annals of the Association of American Geographers*
AAI	Architectural Association of Ireland
CARD	J. T. Gilbert, *Calendar of the Ancient Records of Dublin* (18 vols, 1889 and 1922)
E&P,D	*Environment and Planning D: Society and Space*
FJ	*Freeman's Journal*
GDRM	Greater Dublin Reconstruction Movement
HC	House of Commons
IrB	*Irish Builder*
IrArch	*The Irish Architect and Craftsman*
JHG	*Journal of Historical Geography*
NA	National Archives, Dublin
NLI	National Library of Ireland
OPW	Office of Public Works
OPW	*OPW Annual Reports*
PG	*Political Geography*
PHG	*Progress in Human Geography*
PRIA	*Proceedings of the Royal Irish Academy*
RIAI	Royal Institute of Architects of Ireland
RPDCD	Reports and Printed Documents of the Corporation of Dublin
RDS	Royal Dublin Society
TCD	Trinity College, Dublin
TIBG	*Transactions of the Institute of British Geographers*

Introduction: setting the context

1   See J. Brady and A. Simms (eds), *Dublin through space and time* (Dublin, 2001).

2   See M. Craig, *Dublin 1660–1800* (Dublin, 1980) and E. Sheridan, 'Designing the capital city: Dublin *c.*1660–1810', in Brady and Simms (eds), *Dublin through space and time*, pp. 66–135.

3   The Irish legislature became independent in 1782. However, the British continued to retain executive powers and to manage the Dublin parliament by persuading it to vote for most of the measures they required. The parliament became known as 'Grattan's Parliament' after Henry Grattan, the chief spokesperson of the Irish.

4   See P. Fagan, 'The population of Dublin in the eighteenth century with particular reference to the proportions of Protestants and Catholics', *Eighteenth century Ireland* 6 (1991): 121–56. Fagan's analysis suggests that in 1710 the population of the capital was split *c.*65 per cent Protestant and *c.*35 per cent Catholic, in 1750 the split was almost even at 50 per cent each, and by 1800 the percentage of Protestants had declined to 35 per cent leaving the Catholic population in a majority with 65 per cent.

5   S. J. Connolly, 'Culture, identity and tradition', in B. Graham, (ed.), *In search of Ireland* (London, 1997), pp. 43–63, p. 49.

6   On 1 Jan. 1801 the United Kingdom of Great Britain and Ireland came into being and Ireland ceased to have a parliament. Instead it was to elect 100 members of parliament to the House of Commons in Westminster and contribute a number of peers and bishops to the House of Lords.

7   See M.E. Daly, *Dublin, the deposed capital: A social and economic history, 1860–1914* (Cork, 1985).

8   See J. Prunty, *Dublin slums, 1850–1925* (Dublin, 1998).

9   On the debate surrounding Ireland's status as a colony see for example T. Bartlett, 'What ish my nation? Themes in Irish history, 1550–1850', in T. Bartlett et al. (eds), *Irish studies: a general introduction* (Dublin, 1988), pp. 44–59; L. Kennedy, *Colonialism, religion and nationalism in Ireland* (Belfast, 1996); D. Lloyd, 'Introduction', in D. Lloyd, *Ireland after history* (Cork, 1999), pp. 1–18, S. Howe, *Ireland and empire* (Oxford, 2000).

10   A. J. Christopher, '"The second city of the empire": colonial Dublin, 1911', *JHG* 23, 2 (1997): 151–63.

11   D. Fitzpatrick, *The two Irelands, 1912–1939* (Oxford, 1998), p. 6.

12   *The Irish Times,* 1 Jan. 1900.

13   S. Howe, *Ireland and empire* (Oxford, 2000), p. 68

14   The Home Rule Act was passed on 18 Sept. 1914.

15   The Anglo-Irish Treaty was signed in London in December 1921. The terms of the Treaty provoked a deep split in Irish politics and led in turn to the Irish Civil War of 1922–3.

Chapter 1 Interrogating the cultural landscape

1   S. Daniels and D. Cosgrove, 'Introduction', in D. Cosgrove and S. Daniels (eds), *The iconography of landscape* (Cambridge, 1988), pp. 1–10, p. 1.

2   D. Cosgrove and P. Jackson, 'New directions in cultural geography', *Area* 19, 2 (1987): 95–101, p. 96.

3   D. Mitchell, *Cultural geography: a critical introduction* (Oxford, 2000), p. xiv.

4   See A. Buttimer, *Values in geography,* AAG Commission on College Geography, Resource Paper, no. 24 (Washington DC, 1974); A. Buttimer, 'Reason, rationality and human creativity', *Geografiska Annaler* 61B (1979): 43–9; Y-F. Tuan, 'Place: an experiential perspective', *Geographical Review* 65 (1975): 151–65 and *Topophilia: a study of environmental perception, attitudes and values* (Englewood Cliffs, 1974). See also D. Cosgrove, 'John Ruskin and the geographical imagination', *Geographical Review* 69 (1979): 43–62; S. Daniels, 'Arguments for a humanistic geography', in R. J. Johnston (ed.), *The future of geography* (London, 1985),

pp. 143–58; J. N. Entrikin, 'Contemporary humanism in geography', *AAAG* 66 (1976): 615–32; D. Meinig, 'Geography as an art', *TIBG* 8 (1983): 314–28; D. Meinig (ed.), *The interpretation of ordinary landscapes* (New York, 1979).

5   See K. Anderson and F. Gale (eds), *Cultural geographies* (London, 1999); M. Crang, *Cultural geography* (London, 1998); P. Jackson, *Maps of meaning* (London, 1989); Mitchell, *Cultural geography*; P. Shurmer-Smith and K. Hannam, *Worlds of desire, realms of power* (London, 1994); and P. Shurmer-Smith (ed.), *Doing cultural geography* (London, 2001).

6   Shurmer-Smith (ed.), *Doing cultural geography*, p. 3. On the evolution of the 'new cultural geography' see Cosgrove and Jackson, 'New directions'; D. Cosgrove, 'Geography is everywhere: culture and symbolism in human landscapes', in D. Gregory and R. Walford (eds), *Horizons in human geography* (Basingstoke, 1989), pp. 118–35, J. S. Duncan, 'After the civil war: reconstructing cultural geography as heterotopia', in K. E. Foote et al. (eds), *Re-reading cultural geography* (Austin, 1994), pp. 401–8; K. M. Dunn, 'Cultural geography and cultural policy', *Australian Geographical Studies* 35, 1 (1997): 1–11; P. Jackson, 'Geography and the cultural turn', *Scottish Geographical Magazine* 113, 3 (1997): 186–8, L. L. L. Kong, 'A "new" cultural geography? Debates about invention and reinvention', *Scottish Geographical Magazine* 113, 3 (1997): 177–85; L. McDowell, 'The transformation of cultural geography', in D. Gregory et al. (eds), *Human Geography* (London, 1994), pp. 146–73.

7   C. O. Sauer, 'The morphology of landscape', *University of California Publications in Geography* 2 (1925): 19–54.

8   See P. Wagner and M. Mikesell, *Readings in cultural geography* (Chicago, 1962).

9   K. Anderson and F. Gale, 'Introduction', in Anderson and Gale (eds), *Cultural geographies*, pp. 1–21, p. 4. On the culture concept and the superorganic approach see J. S. Duncan, 'The superorganic in American cultural geography', *AAAG* 70, 2 (1980): 181–98. Although many have charged Sauer with a weak, superorganic conceptualisation of culture, Price and Lewis have argued against this position in their stout defence of the classical school, see M. Price and M. Lewis, 'The reinvention of cultural geography', *AAAG* 83 (1993): 1–17 and the ensuing commentary; D. Cosgrove, 'On the reinvention of cultural geography by Price and Lewis: Commentary'; J. S. Duncan, 'On "The reinvention of cultural geography", by Price and Lewis. Commentary'; P. Jackson, 'Berkeley and beyond: Broadening the horizons of cultural geography', and M. Price and M. Lewis, 'Reply: On reading cultural geography', all in *AAAG* 83 (1993): 515–22.

10   Anderson and Gale, 'Introduction', p. 4.

11   C. Nelson, P. Treichler and L. Grossberg, 'Cultural studies: an introduction', in L. Grossberg, C. Nelson and P. Treichler (eds), *Cultural studies* (New York, 1992), pp. 1–16, p. 5. On the debate surrounding the reconceptualisation of culture see D. Mitchell, 'There's no such thing as culture: towards a reconceptualisation of the idea of culture in geography', *TIBG* 20 (1995): 102–16; P. Jackson, 'The idea of culture: a response to Don Mitchell', D. Cosgrove, 'Ideas and culture: a response to Don Mitchell', J. Duncan and N. Duncan, 'Reconceptualising the idea of culture in geography: a reply to Don Mitchell', D. Mitchell, 'Explanation in cultural geography: a reply to Cosgrove, Jackson and the Duncan's', all in *TIBG* 21, 3 (1996): 572–83.

12   Shurmer-Smith (ed.), *Doing cultural geography*, p 29.

13   See D. Cosgrove, 'Prospect, perspective and the evolution of the landscape idea', *TIBG* 10, 1 (1985): 45–62; D. Cosgrove, *Social formation and symbolic landscape* (London, 1984) and *The Palladian landscape* (Leicester, 1996); M. Williams, 'Historical geography and the concept of landscape', *JHG* 15 (1989): 92–104; R. Muir, 'Landscape: a wasted legacy', *Area* 30 (1998): 263–71; A. R. H. Baker, 'Introduction: on ideology and landscape', in A. R. H. Baker and G. Biger (eds), *Ideology and landscape in historical perspective* (Cambridge, 1993), pp. 1–14.

14   H. Leitner and P. Kang, 'Contested urban landscapes of nationalism: the case of Taipei', *Ecumene* 6, 2 (1996): 214–33, p. 217. See also B. Bender, 'Introduction: landscape – meaning and action', in B. Bender (ed.), *Landscape: politics and perspectives* (Oxford, 1992), pp. 1–18; J. Jacobs, 'The city unbound: qualitative approaches to the city', *Urban Studies* 30 (1993): 827–48.

15   L. Kong, 'Political symbolism of religious building in Singapore', *E&P,D* 11 (1993): 23–45, p. 24. See also L. Kong, 'Negotiating conceptions of "sacred space": a case study of religious buildings in Singapore', *TIBG* 18 (1993): 342–58.

16   J. Duncan and N. Duncan, '(Re)reading the landscape', *E&P,D* 6 (1988): 117–26, p. 118. See also S. Daniels and D. Cosgrove, 'Spectacle and text: landscape metaphors in cultural geography', in J. S. Duncan and D. Ley (eds), *Place/Culture/Representation* (London, 1993), pp. 57–77.

17   Duncan and Duncan, '(Re)reading the landscape', p. 117.

18   J. Duncan, 'Landscape geography', *PHG* 19 (1995): 414–22.

19   See W. Natter and J. P. Jones, 'Signposts toward a poststructuralist geography', in J. P. Jones, W. Natter and T. R. Schatzki (eds), *Postmodern contentions: epochs, politics, space* (New York and London, 1993), pp. 165–203. This essay outlines the nature of the author–text–literary critic/agent–space–geographer triad and provides an historical account of both literary theory and geography and the changing modes of interpretation in the twentieth century that have variously privileged the authorship, the object itself or the interpreter. See especially pp. 171–85.

20   Duncan and Duncan '(Re)reading the landscape', pp. 117–26. See also Duncan's work on the ceremonial city of Kandy in which he demonstrates how the landscape of the ancient capital can be read as a contested space of representation, in J. Duncan, *The City as Text: the politics of landscape interpretation in the Kandyan kingdom* (Cambridge, 1990), also, J. Duncan, 'The power of place in Kandy, Sri Lanka, 1780–1980', in J. A. Agnew and J. S. Duncan (eds), *The Power of Place: bringing together the geographical and sociological imaginations* (Cambridge, 1989), pp. 185–201; 'Representing power, the politics and poetics of urban form in the Kandyan kingdom', in Duncan and Ley (eds), *Place/Culture/Representation*, pp. 232–50. On the pros and cons of the textual metaphor see J. Walton, 'How real(ist) can you get?', *Professional Geographer* 47, 1 (1995): 61–5; D. Mitchell, 'Sticks and stones: the work of landscape', *Professional Geographer* 48 (1996): 94–6; R. Peet, 'Discursive idealism in the "landscape as text school"', *Professional Geographer* 48 (1996): 96–8; J. R. Walton, 'Bridging the divide – a reply to Mitchell and Peet', *Professional Geographer* 48 (1998): 98–100; E. McCann, 'Landscape, texts and the politics of planning', *E&P,D* 15 (1997): 641–61; R. Schein, 'The place of landscape: a conceptual framework for interpreting the American scene', *AAAG* 87 (1997): 660–80.

21   P. Knox, 'The restless urban landscape: economic and socio-cultural change and the transformation of metropolitan Washington', *AAAG* 81 (1991): 181–209. See also D. Ley, 'Cultural/humanistic geography', *PHG* 9 (1985, 9): 415–23. In his analysis of the landscapes of

inner Vancouver, Ley points out that, 'we might identify the landscape as a text, as a cultural form which upon interrogation reveals a human drama of ideas and ideologies, interest groups and power blocs nested within particular social and economic contexts'. Further, 'landscape style is intimately related to the historic swirl of culture, politics, economics, *and* personality in a particular place at a particular time. It is a medium to be read for ideas, practices, interests and contexts constituting the society which created it', see D. Ley, 'Styles of the times: liberal and neo-conservative landscapes in inner Vancouver, 1968–1986', *JHG* 13 (1987): 40–56, p. 41.

22   On symbolism and the urban landscape see for example T. H. Aase, 'Symbolic space: representations of space in geography and anthropology', *Geografiska Annaler* 75, B, 1 (1994): 51–8; J. Eyles and W. Peace, 'Signs and symbols in Hamilton: an iconology of Steeltown', *Geografiska Annaler* 72, B (1990): 73–87; R. Peet, 'A sign taken for history: Daniel Shays' memorial in Petersham, Massachusetts', *AAAG* 86, 1 (1996): 21–43; L. Rowntree and M. Conkey, 'Symbolism and the cultural landscape', *AAAG*, 70, 4 (1980): 459–74.

23   T. J. Barnes and D. Gregory (eds), *Reading human geography: the poetics and politics of inquiry* (London, 1997), p. 508. See Daniels and Cosgrove, 'Introduction', in *The iconography of landscape*; G. Rose, 'Discourse analysis 1: text, intertextuality, context', in G. Rose, *Visual methodologies* (London, 2001), pp. 135–63; E. Panofsky, *Studies in iconology: humanistic themes in the art of the renaissance* (Oxford, 1939) and E. Panofsky, 'Iconography and iconology: an introduction to the study of Renaissance art', in E. Panofsky, *Meaning in the visual arts* (Harmondsworth, 1970), pp. 51–81. The anthropologist Clifford Geertz has adopted the iconographic method with his framework of 'thick' and 'thin' description of cultures, see C. Geertz, 'Thick description: towards an interpretative theory of culture', in C. Geertz, *The interpretation of cultures* (New York, 1973), pp. 3–30.

24   Cosgrove, 'Geography is everywhere', p. 125.

25   P. Gruffudd, 'Remaking Wales: nation building and the geographical imagination, 1925–50', *PG* 14, 3 (1995): 219–39, p. 220.

26   J. Bell, 'Redefining national identity in Uzbekistan', *Ecumene* 6, 2 (1999): 183–207, p. 186.

27   See E. Renan, 'What is a nation?', trans. M. Thom, in H. Bhabha (ed.), *Nation and narration* (London, 1990), pp. 8–22, p. 19.

28   H. Leitner and P. Kang, 'Contested urban landscapes of nationalism: the case of Taipei', *Ecumene* 6, 2 (1996): 214–33, p. 216. On the role of art see D. Ades (ed.), *Art and power: Europe under the Dictators, 1930–45* (London, 1995); I. Golomstock, *Totalitarian art in the Soviet Union, the Third Reich, Fascist Italy and the People's Republic of China* (New York, 1990); B. S. Osborne, 'Warscapes, landscapes, inscapes. France, War and Canadian national identity', in I. S. Black and R. A. Butlin (eds), *Place, culture and identity* (Quebec, 2001), pp. 311–34; B. S. Osborne, 'The iconography of nationhood in Canadian art', in Cosgrove and Daniels (eds), *Iconography of landscape*, pp. 162–78; B. S. Osborne, 'Interpreting a nation's identity: artists as creators of national consciousness', in Baker and Biger (eds), *Ideology and landscape*, pp. 230–54; S. Daniels, *Fields of vision: landscape imagery and national identity in England and the United States* (Princeton, 1993).

29   Renan, 'What is a nation?', p. 11. For discussion of the notion of heritage as cultural capital and how this relates to legitimation theory and dominant ideology see G. J. Ashworth, *On tragedy and Renaissance* (Groningen, 1993), especially ch. 10; G. J. Ashworth and P. Howard,

'Heritage and identity at different spatial scales', in G. J. Ashworth and P. Howard, *European heritage planning and management* (Exeter, 1999).

30   M. Azaryahu, 'German reunification and the power of street names', *PG* 16, 6 (1997): 479–93, p. 480.

31   See the range of essays in F. Driver and D. Gilbert, (eds), *Imperial cities: landscape, display and identity* (Manchester, 1999), and F. Driver and D. Gilbert, 'Imperial cities, overlapping territories, intertwined histories', in *ibid.*, pp. 1–17.

32   B. Anderson, *Imagined communities* (London, 1983). See also E. Hobsbawm and T. Ranger, *The invention of tradition* (Cambridge, 1983), especially ch. 3. On the overlapping relationships between heritage, memory and identity see J. R. Gillis (ed), *Commemorations. The politics of national identity* (Princeton, 1994); B. Graham, G. J. Ashworth and J. E. Tunbridge, *A geography of heritage* (London, 2000), especially Part 1; N. Johnson, 'Historical geographies of the present', in B. Graham and C. Nash (eds), *Modern historical geographies* (Harlow, 2000), pp. 251–72; D. Lowenthal, *The heritage crusade and the spoils of history* (London, 1996); P. Nora, 'Between memory and history', les lieux des mémoires', *Representations* 26 (1989): 10–18; E. Said, 'Invention, memory, and place', *Critical Inquiry* 26 (2000): 175–92; R. Samuel, *Theatres of memory; vol. 1, past and present in contemporary culture* (London, 1994); S. Schama, *Landscape and memory* (London, 1995); 374–402; J. E. Tunbridge and G. J. Ashworth, *Dissonant heritage: the management of the past as a resource in conflict* (Chichester, 1996).

33   Graham et al., *A geography of heritage*, p. 18. See also B. Graham, 'The past in place: historical geographies of identity', in Graham and Nash (eds), *Modern historical geographies*, pp. 70–99.

34   J. R. Gillis, 'Memory and identity: the history of a relationship', in Gillis (ed.), *Commemorations*, pp. 3–24, p. 4.

35   D. Lowenthal, 'Identity, heritage, and history', in Gillis (ed.), *Commemorations*, pp. 41–60, p. 43. See also, R. Handler, 'Is 'Identity' a useful cross-cultural concept?' in Gillis (ed.), *Commemorations*, pp. 27–40.

36   B. Graham, G. J. Ashworth and J. E. Tunbridge, *A geography of heritage* (London, 2000), p. 35.

37   J. Bell, 'Redefining national identity in Uzbekistan', *Ecumene* 6, 2 (1999): 183–207, p. 186.

38   This has been conceptualised through the notion of dissonance, that is, 'a discordance or a lack of agreement and consistency in the definition of heritage', which comes to prominence when groups within multi-cultural societies often do not identify with the dominant, officially designated heritage, see J. E. Tunbridge, 'The question of heritage in European cultural conflict', in B. Graham (ed.), *Modern Europe: place, culture, identity* (London, 1998), pp. 236–60.

39   See B. S. A. Yeoh, 'Historical geographies of the colonised world', in B. Graham and C. Nash (eds), *Modern historical geographies*, pp. 146–66, p. 146.

40   *Ibid.*, p. 147. See also B. S. A. Yeoh, 'Postcolonial cities', *PHG* 25, 3 (2001): 456–68.

41   K. E. Foote, A. Tóth and A. Árvay, 'Hungary after 1989: inscribing a new past on place', *Geographical Review* 90, 3 (2000): 301–34. See also R. Argenbright, 'Remaking Moscow: new places, new selves', *Geographical Review* 89, 1 (1999): 1–22; B. Ladd, 'East Berlin political monuments in the late German Democratic Republic: finding a place for Marx and Engels', *Journal of Contemporary History* 37, 1 (2002): 91–104.

42 H. Lefebvre, *The production of space* (Oxford, 1991), p. 54.

43 N. Johnson, 'Cast in stone: monuments, geography, and nationalism,' *E&P,D* 13 (1995): 51–65, p. 52.

44 G. Owens, 'Nationalist monuments in Ireland, 1870–1914: symbolism and ritual', in R. Gillespie and B. P. Kennedy (eds), *Ireland: art into history* (Dublin, 1994), pp. 103–17, p. 103. See also S. Levinson, *Written in stone: public monuments in changing societies* (Durham, N. C., 1998); S. Michalski, *Public monuments: art in political bondage, 1870–1997* (London, 1998).

45 M. Agulhon, 'La statuomanie et l'histoire', *Ethnologie française* 3/4 (1978): 145–72, quoted in Owens, 'Nationalist monuments in Ireland', p. 103.

46 Governments made use not only of monuments but also of an array of symbols and ritualistic devices to mobilise mass support. Public festivals, national emblems, banners and patriotic songs also had a role to play in giving expression to power. See M. Agulhon, *Marianne into battle: republican imagery and symbolism in France, 1789–1880* (Cambridge, 1981); A. Boime, *Hollow icons: the politics of sculpture in nineteenth century France* (Kent, Ohio and London, 1987); W. Cohen, 'Symbols of power: statues in nineteenth century provincial France', *Comparative Studies in History and Society* 31 (1989): 491–513; G. L. Mosse, *The nationalisation of the masses: political symbolism and mass movements in Germany from the Napoleonic Wars through the Third Reich* (New York, 1975); E. Kamenka, 'Caesarism, circuses and monuments', in G. L. Mosse (ed.), *Masses and man: nationalist and fascist perceptions of reality* (New York, 1980), pp. 104–18; E. Hobsbawm, Foreword, in D. Ades, T. Benton, D. Elliot, I. Boyd-White (eds), *Art and power: Europe under the dictators* (London, 1995); E. Hobsbawm, 'Mass producing traditions: Europe, 1870–1914', in E. Hobsbawm and T. Ranger (eds), *The invention of tradition* (Cambridge, 1983), pp. 263–309; M. North, 'The public as sculpture: from heavenly city to mass ornament', in W. J. T. Mitchell (ed.), *Art and the public sphere* (Chicago, 1990), pp. 9–28; A. J. Lerner, 'The nineteenth century monument and the embodiment of national time', in M. Ringrose and A. J. Lerner (eds), *Reimagining the nation* (Buckingham, 1993), pp. 176–96.

47 Lerner, 'The nineteenth century monument', p. 178.

48 C. W. J. Withers, 'Place, memory, monument: memorialising the past in contemporary Highland Scotland', *Ecumene* 3 (1996): 325–44, p. 327.

49 T. Edensor, 'National identity and the politics of memory: remembering Bruce and Wallace in symbolic space', *E&P,D* 29 (1997): 175–94.

50 L. Sandercock, *Towards cosmopolis* (New York, 1988), p. 207.

51 See D. I. Kertzer, *Ritual, politics and power* (New Haven and London, 1988); L. Jakubowska, 'Political drama in Poland: the use of national symbols', *Anthropology Today* 6 (1990): 10–13.

52 These actions echoed those of decades earlier when the public statues dedicated to members of the Tsars' family were also summarily removed. In his book *Architecture and ideology in Eastern Europe during the Stalin era*, Anders Åman writes at length of the monuments erected in the cold war era and their symbolic meaning and significance both when erected and how this meaning has evolved through time. See also D. Sidorov, 'National monumentalisation and the politics of scale', *Annals of the Association of American Geographers* 90, 3 (2000): 548–72.

53   P. Parkhurst Ferguson, 'Reading city streets', *French Review* 61, 3 (1988): 386–97, p. 386.

54   M. Azaryahu, 'The power of commemorative street names', *E&P,D* 14 (1996): 311–30, p. 311.

55   M. Azaryahu, 'German reunification and the politics of street names: the case of Berlin', *PG* 16, 6 (1997): 479–93, p. 481. See also See also M. Azaryahu, 'Street names and political identity: the case of East Berlin', *Journal of Contemporary History* 21 (1986): 581–604; M. Azaryahu, 'Renaming the past: changes in 'city text' in Germany and Austria, 1945–1947', *History and Memory* 22 (1990): 32–53; M. Azaryahu, 'The spontaneous formation of memorial space: the case of Kikar Rabin, Tel Aviv', *Area* 28, 4 (1996): 501–13; M. Azaryahu, 'McDonald's or Golani Junction? A case of a contested place in Israel', *Professional Geographer* 51, 4 (1999): 481–92.

56   K. Palonen, 'Reading street names politically', K. Palonen and T. Parviko, (eds), *Reading the political: exploring the margins of politics* (Tampere, 1993), pp. 103–21.

57   J. R. Short, *The urban order* (Oxford, 1996), p. 299. Events in Germany echo this process where in Apr. of 1990 the residents of the city of Karl-Marx-Stadt decided to change its name back to Chemnitz. Thus, after 37 years the old name of the city was restored in a gesture which symbolised the dismantling of the communist regime.

58   J. R. Short, *The urban order* (Oxford, 1996), p. 425.

59   J. N. Entrikin, *The betweeness of place* (Baltimore, 1991), pp. 55–6.

60   J. Kenny, 'Portland's comprehensive plan as text: the Fred Meyer case and the politics of reading', in T. J. Barnes and J. S. Duncan (eds), *Writing worlds: discourse, text and metaphor in the representation of landscape* (London, 1992), pp. 176–92, p. 176. See also G. Macdonald, 'Indonesia's *Medan Medaka*: national identity and the built environment', *Antipode* 27, 3 (1995): 270–93.

61   Cosgrove, 'Geography is everywhere', pp. 128–9.

62   For an account of the links between planning and national identity, along with architecture, see N. Al Sayyad (ed.), *Forms of dominance: on the architecture and urbanism of the colonial enterprise* (Aldershot, 1992); L. J. Vale, *Architecture, power and national identity* (New Haven and London, 1992). See also A. E. J. Morris, *History of urban form before the industrial revolutions* (London, 1994); P. Hall, *Cities of tomorrow* (Oxford, 1988); P. Hall, *Cities in civilisation* (London, 1998); T. Hall, *Planning Europe's cities, aspects of nineteenth century urban development* (London, 1997); S. Kostoff, *The city shaped* (London, 1991). Useful articles on a similar theme can be found in the journal *Planning Perspectives*. See, for example, M. Cavalcanti, 'Urban reconstruction and autocratic regimes: Ceausescu's Bucharest in its historic context', *Planning Perspectives* 12 (1997): 71–109; V. Hastaglou-Martinidis, K. Kaflouka and N. Papanichos, 'Urban modernisation and national renaissance; town planning in 19th century Greece', *Planning Perspectives*, 8 (1993): 427–69; A. Yerolympos, 'A new city for a new state: city planning and the formation of national identity in the Balkans', *Planning Perspectives* 8 (1993): 233–57; see also G. A. Myers, 'Making the socialist city of Zanzibar', *Geographical Review* 84, 2 (1994): 451–62.

63   C. Jencks, *Architecture today* (New York, 1982), p. 178.

64   See K. Dovey, 'Corporate towers and symbolic capital', *Environment and Planning B: Planning and Design* 19 (1992): 173–88.

65   Department of Arts, Culture and the Gaeltacht, *Developing a government policy on architecture: a proposed framework and discussion of issues* (Dublin, 1996), p. 69.

66   J. Goss, 'The built environment and social theory: towards an architectural geography', *Professional Geographer* 40, 4 (1988): 392–403, p. 392.

67   See also L. Bondi, 'Gender symbols and urban landscapes', *PHG* 16 (2) 1992: 157–70. In her gendered reading of the architecture of the urban landscape she points out that in suburban residential architecture, 'feminine coding operates principally through associations with nurturance, domesticity and so on, but again, beliefs about the distinctiveness of women's bodies are at work in the use of curves, and of nooks and crannies . . . suburbia itself resonates with assumptions about the beneficence of nuclear family living, "complementary gender roles and heterosexuality"', p. 160.

68   L. Mumford, *The culture of cities* (New York, 1983), p. 403, cited in P. L. Knox, 'The social production of the built environment. Architects, architecture and the post-modern city', *PHG* 11 (1987): 354–78.

69   See Daniels, *Fields of vision*.

70   S. Seymour, 'Historical geographies of landscape', in Graham and Nash (eds), *Modern historical geographies*, pp. 193–217, p. 194.

71   This research reflects recent theoretical shifts in the social sciences and humanities away from positivist models of research towards an acknowledgement of the inherently situated nature of research and the highly political nature of representation, see Graham and Nash, 'Preface', *Modern historical geographies*, p. xiii.

72   B. S. A. Yeoh, 'Street names in colonial Singapore', *Geographical Review* 82, 3 (1992): 313–23, and B. S. A. Yeoh, 'Street-naming and nation-building: toponymic inscriptions of nationhood in Singapore', *Area* 28, 3 (1996): 298–307.

73   Yeoh, 'Street-naming and nation-building', p. 300.

74   *Ibid.*

75   G. A. Myers, 'Naming and placing the other: power and the urban landscape in Zanzibar', *Tijdschrift voor Economische en Sociale Geografie* 87, 3 (1996): 237–46, p. 241.

76   S. B. Cohen and N. Kliot, 'Place-names in Israel's ideological struggle over the administered territories', *AAAG* 82, 4 (1992): 653–80.

77   M. Azaryahu, 'The power of commemorative street names', *E&P, D* 14, 3 (1996): 311–30.

78   M. Azaryahu, 'German reunification and the politics of street names: the case of East Berlin', *PG* 16, 6 (1997): 479–93.

79   *Ibid.*

80   A. Pred, 'Capitalisms, crises, and cultures II: notes on local transformation and everyday struggles', in A. Pred and M. Watts (eds), *Reworking modernity: capitalisms and symbolic discontent* (New Brunswick, NJ, 1992), pp. 106–17, 139–41.

81   *Ibid.*, pp. 139–41.

82   L. D. Berg and R. A. Kearns, 'Naming as norming: "race", gender, and the identity politics of naming places in Aoteraoa/New Zealand', *E&P, D* 14 (1996): 99–122.

83   'Pakeha' is the term used to denote New Zealand-born people of European descent.

84   In his work on late nineteenth century rural protest in contemporary Highland Scotland, Withers has explored the meaning embedded in commemorative monuments with particular reference to 'the socially active and constitutive role of memory in giving those events present

meaning', C. Withers, 'Place, memory, monument: memorialising the past in contemporary Highland Scotland', *Ecumene* 3, 3 (1996): 325–44, p. 326.

85   M. Heffernan, 'For ever England: the Western Front and the politics of remembrance in Britain', *Ecumene* 2, 3 (1995): 293–324, p. 295.

86   B. Graham, 'No place of mind: contested Protestant representations of Ulster', *Ecumene* 1 (1994): 257–82.

87   N. Johnson, 'Cast in stone, monuments, geography and nationalism', *E&P,D* 13 (1995): 51–65. She has also highlighted the particular significance of the politics of commemoration with regard to the commemoration of rebellion and how statues were used to celebrate Ireland's attempt to achieve independence from Britain, see N. Johnson, 'Sculpting heroic histories: celebrating the centenary of the 1798 rebellion in Ireland', *TIBG* 19 (1994): 78–93.

88   J. Duncan, *The city as text: the politics of landscape interpretation in the Kandyan kingdom* (Cambridge, 1990).

89   D. Atkinson and D. Cosgrove, 'Urban rhetoric and embodied identities: city, nation and empire at the Vittorio Emanuele II monument in Rome, 1870–1945', *AAAG* 88, 1 (1998): 28–49, p. 28.

90   B. Osborne, 'Constructing landscapes of power: the George Étienne Cartier monument, Montreal', *JHG* 24, 4 (1998): 431–58.

91   R. Peet, 'A sign taken for history: Daniel Shays' memorial in Petersham, Massachusetts', *AAAG* 86, 1 (1996): 21–43.

92   D. Harvey, 'Monument and myth', *AAAG* 69 (1979): 362–81, p. 381. See also D. Harvey, 'Paris, 1850–1870', in D. Harvey (ed.), *Consciousness and the urban experience* (Baltimore, 1985).

93   P. Shurmer-Smith and K. Hannam, 'Monuments and spectacles', in *Worlds of desire*, pp. 198–214, p. 203.

94   M. Domosh, 'Corporate cultures and the modern landscape of New York city', in K. Anderson and F. Gale (eds), *Inventing places* (Melbourne, 1995), pp. 73–86. See also M. Domosh, 'The symbolism of the skyscraper: case studies of New York's first tall buildings', *Journal of Urban History* 14 (1988): 321–45; 'A method for interpreting landscape: a case study of the New York world building', *Area* 21 (1989): 347–53; 'New York's first skyscrapers: conflict in design of the American commercial landscape', *Landscape* 30 (1989): 34–8; 'The symbolism of the skyscraper: case studies of New York's first tall buildings', in Foote, et al. (eds), *Re-reading cultural geography*, pp. 48–63.

95   I. S. Black, 'Symbolic capital: the London and Westminster Bank headquarters, 1836–38', *Landscape Research* 21, 1 (1996): 55–72.

96   P. Woolf, 'Symbol of the second empire: cultural politics and the Paris Opera House' in Cosgrove and Daniels (eds), *The iconography of landscape*, pp. 214–35.

97   See Al Sayyad, *Forms of dominance.*

98   T. Haarni, 'Modern Art and revenge of the military hero: new planning practices and the geography of resistance in central Helsinki', *Nordisk Samhällsgeografisk Tidskrift* 20 (1995): 116–30, p. 123. See also H. Lorimer, 'Sites of authenticity: Scotland's new parliament and official representations of the nation', in D. C. Harvey, R. Jones, N. McInroy and C. Milligan (eds), *Celtic geographies: old culture, new times* (London, 2002), pp. 91–108.

99   S. Lewandowski, 'The built environment and cultural symbolism in post-colonial Madras', in J. Agnew, J. Mercer and D. Sopher (eds), *The city in cultural context* (London, 1984), pp. 237–54.

100 Vale, *Architecture, power and national identity.*

Chapter 2   Representing empire

1   MCG, *The monuments of Dublin: a poem* (Dublin and Belfast, 1865), pp. 1–20, p. 1.

2   W.M. Thackeray, *The Irish sketchbook* (Dublin, 1842), p. 16.

3   *IrB*, 1 Nov. 1864, p. 216.

4   MCG, *Monuments of Dublin*, p. 3.

5   This position was safeguarded a year later at the Battle of Aughrim, when the Irish Catholic forces were decisively defeated. Thereafter, the Irish Parliament remained in the firm control of the Anglican landed classes, while William III was virtually deified by Protestant Ireland.

6   J. Kelly, '"The glorious and immortal": memory, commemoration and Protestant identity in Ireland, 1660–1800', *PRIA* 54, c, 2 (1992): 26–52, p. 30.

7   See J. T. Gilbert, *History of the city of Dublin* (Dublin and London, 1903), pp. 29–30.

8   Further evidence of Protestant Dublin's 'deification' of King William is evident in the fact that when the Houses of Parliament were opened in 1735, the chamber of the House of Lords was decorated with a specially commissioned tapestry depicting the Battle of the Boyne. Also, an obelisk was erected in 1736 on the banks of the Boyne at Oldbridge in Co. Meath, opposite the scene of some of the fiercest fighting of the Battle of the Boyne. This monument was given an inscription that pointed to the 'glorious memory of King William III, who crossed the Boyne near this place to attack James II at the head of a Popish army advantageously placed on the south side of it, and did on that day, by a successful battle secure to us and our posterity, our liberty, laws and religion'. Another memorial was erected in Boyle, Co. Roscommon. The icon of King William also became popular with Orange clubs formed in the eighteenth century, see B. Loftus, *Mirrors: William III and modern Ireland* (Down, 1990).

9   *CARD*, VI, p. 232. The preface also records that in Apr. of 1700, 'Henry Glegg and John Moore, of Dublin, under authority from the Lord Mayor and Sheriffs, contracted at London with Grinling Gibbons for the execution of the work for eight hundred pounds, see *CARD*, VI, p. 235. The corn market area had initially been suggested as a possible location but this was later rejected in favour of the College Green site.

10   *CARD*, VI, p. 237.

11   Preface to *CARD*, VI, pp. vii–xiv.

12   Gilbert, *History of Dublin*, p. 27.

13   *Ibid.*, p. 35.

14   *Ibid.*, p. 30. As Craig notes, 'it was subjected to repeated defacement and abuse, daubed with filth, tarred, robbed of its sceptre, even beheaded', M. Craig, *Dublin, 1660–1860* (Dublin, 1980), p. 76.

15   Gilbert, *History of Dublin*, p. 35.

16   *Ibid.*, p. 31.

17   *Ibid.*

18 *Ibid.*, p. 32. The latter part of the sentence was later revoked and their fines reduced to five shillings.

19 *Ibid.*

20 *Ibid.*, p. 44.

21 On 12 July 1822 a group of Orangemen marched in procession to the statue and later that evening the four lamps surrounding the monument were demolished, the orange insignia was torn from the statue and an Orangeman was attacked. They managed to regain control aided by a detachment of police and Yeomanry and forced all passers by to take off their hats to the statue. *Ibid.*, pp. 49–50.

22 For example, 'On the night of Thursday, the 7th of April, at a few minutes past twelve o'clock, a light appeared suddenly on the northern side of the statue, and immediately afterwards the figure of the king was blown several feet into the air with a deafening explosion, extinguishing the lamps on College Green and its vicinity . . . its legs and arms were broken, and its head completely defaced by the fall; the horse was also injured and shattered in several places'. *Ibid.*, pp. 49–50.

23 *FJ*, 5 Sept. 1882.

24 Given its prominent location opposite the building that once housed Dublin's independent parliament, it was perhaps inevitable that in the decades after independence the statue would be removed, see chapter 7.

25 *CARD*, vii, pp. 49–50, vii–viii. See also A. Kelly, 'Van Nost's equestrian statue of George I', *Irish Arts Review* 11 (1995): 103–7.

26 *CARD*, vii, p. 64.

27 *Ibid.*, p. 49.

28 *Ibid.*, pp. 75 , 157–8, 170, 186–7, 195–6.

29 *Ibid.*, p. ix.

30 *CARD*, xiv, p. 271.

31 *Ibid.* This idea was again put forward in 1806, see *CARD*, xv, 1806, p. 489.

32 Quoted in A. Kelly, 'Van Nost's equestrian statue', p. 104.

33 Thackeray, *Irish sketchbook*, p. 16.

34 Entitled, 'The Beau Walk in Stephen's Green', from Thomas Newburgh, *Essays poetical, moral and critical* (Dublin, 1769), quoted in E. McParland, 'A note on George II and St Stephen's Green', *Eighteenth Century Ireland* 2 (1987): 187–95, p. 187. See also, F. E. Dixon, 'The portrait statues of Dublin', *Dublin Historical Record* 7, 4 (1945): 155–60.

35 *CARD*, x, p. 22.

36 *Ibid.*, pp. 83–4. It was also initially suggested that the statue be erected on a new square proposed to be laid out in front of Dublin Castle, see McParland, 'Note on George II', p. 189. He notes that the site then chosen was the centre of St Stephen's Green, aligned with Dawson Street. In April 1756 it was rumoured that eight pathways were to radiate from the statue, hence in John Rocque's initial engraving the eight walks were put in. These were later removed although vestiges remain just detectable on his map of 1756. Eventually the statue was erected aligned with both York and Dawson Streets with no pathway approach apparent.

37 Quoted in *IrB*, 15 Mar. 1878, p. 81.

38 *Ibid.*

39 MCG, *Monuments of Dublin*, pp. 7–8.

40 P. Henchy, 'Nelson's Pillar', *Dublin Historical Record* 10 (1948–9): 53–63. See also *A description of the pillar, with a list of subscribers* (Dublin, 1846) and W. Bolger and B. Share, *And Nelson on his pillar* (Dublin, 1966) for accounts of the circumstances surrounding the erection of the memorial.

41 Henchy, 'Nelson's Pillar', p. 55.

42 *FJ*, 16 Feb. 1808.

43 Cited in *IrB*, 30 June 1923, p. 497.

44 *FJ*, 16 Feb. 1808.

45 *Ibid.*

46 *IrB*, 7 June 1930, p. 308 and 28 Dec. 1946, p. 856.

47 *The Irish Magazine*, Sept. 1809, quoted in P. Henchy, 'Nelson's Pillar', *Dublin Historical Record* 10 (1948–9): 53–63, 59–60.

48 J. Warburton, J. Whitelaw and R.A. Walsh, *A history of the city of Dublin, from the earliest accounts to the present time* (London, 1818), p. 1100.

49 *IrB*, 15 June 1876, p. 179.

50 *IrB*, 15 May 1876, p. 149.

51 HC, Nelson's Pillar (Dublin) Bill, 13 Feb. 1891.

52 *Ibid.*

53 *Ibid.*

54 *Ibid.*

55 *Ibid.*

56 MCG, *Monuments of Dublin*, p. 8.

57 In parliament he had antagonised the more conservative elements of the Tory party by forcing the passage of the Catholic Emancipation Act in 1829.

58 All of the original committee meeting proceedings were destroyed in a fire around 1825. Documents on the testimonial were gathered into a single volume by General Larcom which can now be accessed in NLI. See P. F. Garnett, 'The Wellington testimonial', *Dublin Historical Record*, 1952, pp. 48–61 for a useful summary of these papers.

59 J. McGregor, *Picture of Dublin* (Dublin, 1821), p. 301.

60 *Ibid.*

61 Six designs were received, two of them for obelisks, three for columns and one in the design of a temple.

62 Quoted in McGregor, *Picture of Dublin*, pp. 301–2.

Chapter 3 Contested identities and the monumental landscape

1 Review of J. T. Gilbert, *History of the city of Dublin* in *Dublin University Magazine*, Mar. 1856, pp. 320–3, p. 321.

2 *The Nation*, 27 May 1843, p. 523. The Duke of Wellington was Irish born, but his monument was not completed until 1861.

3 Both statues were sculpted by the Dublin-born sculptor, John Henry Foley.

4 It was removed in 1959 having fallen into ruin.

5   M. Craig, *Dublin, 1660–1860* (Dublin, 1980), p. 104.

6   MCG, *The monuments of Dublin: a poem* (Dublin and Belfast, 1865), pp. 11–17.

7   *FJ*, 8 Aug. 1864.

8   See W. Stokes, *The Life and labours in art and archaeology of George Petrie* (London, 1868) for an account of the Glasnevin monument, especially pp. 369–72, and appendix, pp. 433–37. See also *IrB*, 1 Jan. 1861, p. 388, 1 Oct. 1862, p. 254, and Glasnevin Cemeteries Group, *Glasnevin cemetery: an historic walk* (Dublin, 1997), p. 6.

9   Dublin City Archives hold the records of the O'Connell monument committee, see especially MS Ch 6/1 and 6/2. See also H. Potterton, *The O'Connell Monument* (Cork, 1973).

10   For over twenty years Canon John O'Hanlon acted as honorary secretary of the O'Connell monument committee. The Catholic Church was a strong supporter of the project to erect a national monument and O'Hanlon played a pivotal role in fund-raising and received donations on behalf of the committee, as well as carrying out its correspondence. He also wrote up the detailed *Report of the O'Connell monument committee*, published in 1888.

11   *FJ*, 9 Aug. 1864.

12   *IrB*, 1 Sept. 1862, pp. 216–17. See P. Murphy, 'John Henry Foley's O'Connell monument', *Irish Arts Review* II (1995): 155–6

13   *IrB*, 15 Mar. 1862 and 1 Sept. 1862.

14   *FJ*, 9 Aug. 1864, p. 7.

15   *Ibid.*

16   *IrB*, 15 Oct. 1863, p. 167.

17   *IrB*, 1 Dec. 1863, p. 192.

18   *IrB*, 1 July 1864, p. 125. See also the following articles in *IrB*: 'Native talent vs. absenteeism', 15 July 1864, p. 135; 'The O'Connell national monument and art, resident and non-resident', 15 Aug. 1864, p. 156; 'The O'Connell monument', 1 Oct. 1864, p. 196; 'How native sculptors are encouraged', *IrB*, 1 Aug. 1872, p. 203.

19   *IrB*, 15 Jan. 1865, p. 18 and 1 Feb. 1865, p. 30.

20   *IrB*, 15 May 1865, pp. 121–2 and 129–39.

21   *IrB*, 1 June 1865, p. 143.

22   *IrB*, 15 Apr. 1866, p. 95.

23   *IrB*, 1 Aug. 1866, p. 190 and 15 Aug. 1866, p. 207.

24   *IrB*, 15 Aug. 1871, p. 205.

25   *Ibid.*

26   Cited in *IrB*, 15 Aug.1871, p. 216.

27   *FJ*, 9 Aug. 1882.

28   *FJ*, 14 Aug. 1882.

29   *Ibid.*

30   *The Irish Times*, 27 Dec. 1870.

31   He had played a key role in the 1848 rising at Ballingarry, Co. Tipperary after which he was transported to Tasmania. His sentence was commuted and in 1856 he returned to Ireland. His contributions to *The Nation* newspaper made clear that he no longer held the extremist views he had earlier in his life and he attacked the extremism of the IRB.

32   *IrB*, 15 Oct. 1867, p. 276.

33   See P. Murphy, 'Thomas Farrell, sculptor', *Irish Arts Review* 9 (1993): 196–207. Farrell along with John Henry Foley was among the most prominent of Dublin's sculptors in the latter half of the nineteenth century, who between them accounted for many of the city's public monuments. He was also to sculpt the memorials dedicated to Sir John Gray (1879), Alexander MacDonnell (1878) and Sir Robert Stewart (1898).

34   *IrB*, 1 Oct. 1867, p. 262.

35   *Ibid.* Dublin Corporation granted permission for this site in 1867. It is noteworthy that in their memorandum to the Council in 1867, when requesting permission, the committee made reference to the fact that 'Not far off, Goldsmith and Burke already stand, and soon we hope, Burke and O'Connell will follow', as if to make plain their patriotic intentions.

36   In 1909 a resolution was passed that, 'the inscription on the O'Brien monument between D'Olier Street and O'Connell bridge be so altered as to better identify it with the name of William Smith O'Brien', Minutes, 1909, no. 549. The following year a committee of the whole house recommended that an inscription in Irish be placed on the monument. See RPDCD, 1909, vol. III, no. 286, p. 1087 and Minutes, 1910, no. 103.

37   *The Irish Times*, 27 Dec. 1870. See J. Morisy, *A wreath for the O'Brien statue* (Dublin, 1870).

38   *The Irish Times*, 27 Dec. 1870.

39   *FJ*, 25 June 1879. The monument did not turn out entirely as had been planned. Initially, it had been designed with a representation of Ireland, complete with harp, on the right hand side of the pedestal and incorporated broken fetters to represent the legislative and social wrongs from which the country had been rescued. There was also to have been a figure of patriotism. The necessity of erecting the monument without delay, however, resulted in the statue featuring the figure of Gray alone.

40   Minutes, 1890, nos 125, 154, 203 and RPDCD, 1890, vol. II, no. 62.

41   *Irish Independent*, 8 Aug. 1898.

42   Minutes, 1898, no. 287.

43   The statue was sculpted by John Hughes who was to find fame on Dublin's streets with the statue of Queen Victoria erected in 1908. A portrait figure, it represented Gladstone accompanied by the allegorical statues of Erin, Classical Learning, Finance and Eloquence. A further attempt to have the statue erected in Dublin was made in 1924 when the Phoenix Park was suggested as a possible location. The government declined the offer and the memorial committee offered it instead to the trustees of St Deinol's Library at Hawarden where it was erected in Apr. 1925.

44   *FJ*, 9 Oct. 1899. See the Parnell Monument Committee records in the Redmond papers (NLI, MS 15238), also T. O'Keefe, 'The art and politics of the Parnell monument', *Éire-Ireland* 19 (1984): 6–25; S. Rothery, 'Parnell monument: Ireland and the Beaux Arts', *Irish Arts Review* 4 (1987): 55–7 and N. Johnson, 'Cast in stone: monuments, geography, and nationalism', *E&P,D*, 13 (1995): 51–65, p. 52.

45   *FJ*, 2 Oct. 1911, p. 5.

46   *Ibid.*

47   *Ibid.*

48   *Ibid.*

49   *Ibid.*

50   *Ibid.*

51   See *The origins, progress and completion of the Grattan statue complied by the Hon. John P. Vereker, 1881* (TCD, MS 1703), also *IrB*, 1 July 1871, p. 162, 15 Oct. 1874, pp. 280–1, 1 Dec. 1875, p. 339, 15 Jan. 1876, pp. 19 and 27.

52   *IrB*, 15 Jan. 1876, p. 27.

53   *The Irish Times*, 7 Jan. 1876.

54   *Ibid.*

55   *Ibid.*

56   T. O'Keefe, 'The 1898 efforts to celebrate the United Irishmen', *Éire-Ireland* 19 (1988): 51–73, pp. 52–3. See also *FJ* and *The Irish Times*, June 1897 and M. J. F. McCarthy, *Five years in Ireland, 1895–1900* (London, 1901).

57   See also N. Johnson, 'Sculpting heroic histories: celebrating the centenary of the 1798 rebellion', *TIBG* 19 (1994): 78–93; G. Owens, 'Nationalist monuments in Ireland, 1870–1914: symbolism and ritual', in R. Gillespie and B. P. Kennedy (eds), *Ireland: art into history* (Dublin, 1994), pp. 103–17, and J. Turpin, 'Oliver Sheppard's 1798 Memorials', *Irish Arts Review* 7 (1990–1): 71–80; S. Paseta, '1798 in 1898: the politics of commemoration', *Irish Review* 22 (1998): 46–53; T. O'Keefe, 'The 1898 efforts to celebrate the United Irishmen: the '98 centennial', *Éire-Ireland* 23 (1988): 51–73; '"Who fears to speak of '98?" The rhetoric and rituals of the United Irishmen Centennial 1898', *Éire-Ireland* 28 (1992): 67–91; R. Foster, 'Remembering 1798', in R. F. Foster, *The Irish story* (London, 2001), pp. 211–34.

58   For further detail on the 1898 commemoration of 1798 see T. O'Keefe, 'The 1898 efforts', and T. O'Keefe, 'Who fears to speak of '98?'.

59   *FJ*, 16 Aug.1898, p. 5. NA holds a number of files of general newspaper coverage of the event, see CBS File 17, O25/s and general newspaper coverage, 16–20 Aug. 1898. *Weekly Independent*, 17 Mar. 1898, p. 9; *Leinster Leader*, 1 Sept. 1800, p. 7; *Daily Independent*, 21 June 1898; *The Times*, 16 Aug. 1898.

60   See *The Irish News, Belfast Morning News*, 11 Aug. 1898; *FJ*, 16 Aug. 1898.

61   See *FJ*, 16 Aug. 1898.

62   *Ibid.*

63   *Ibid.*

64   *Weekly Independent*, 30 July 1898, quoted in G. Owens, 'Nationalist monuments in Ireland, 1870–1914: symbolism and ritual', in Gillespie and Kennedy (eds), *Ireland: art into history*, p. 115. The monument was never actually erected but in 1967 another statue was dedicated at the northeast corner of St Stephen's Green.

65   *IrArch*, 1 Apr. 1911, p. 133.

66   *FJ*, 24 May 1909, p. 4.

67   *Ibid.*

68   *IrArch*, 1 Apr. 1911, p. 133.

69   C. T. M'Cready, *Dublin street names* (Dublin, 1892, reprinted 1987), p. 123.

70   *IrB*, 1 Jan. 1862, p. 7.

71   *Ibid.*

72   *IrB*, 15 Jan. 1862, p. 20.

73   *IrB*, 15 Mar. 1862, p. 69.

74   *IrB*, 1 Jan. 1865, p. 12.

75   When the committee told him that they wished to erect a statue in his honour he requested that 'I should like it to be placed as near as possible to the card-room of my club', *The Irish Times*, 27 Aug. 1958. When the statue of the Duke was bombed in 1958 his club was one of the buildings which suffered from the blast of the explosion.

76   *IrB*, 1 Sept. 1866, p. 215.

77   J. Hill, *Irish public sculpture* (Dublin, 1998), p. 102.

78   *The Irish Times*, 3 May 1870. On a slab at the opposite side the words, 'erected by public subscription' were inscribed.

79   *IrB*, 16 Jan. 1862, p. 16.

80   By Apr. 1862 £2,000 had been raised, see *IrB*, 1 Apr. 1862, and 15 Jan. 1863, p. 3.

81   *IrB*, 1 Apr. 1862. See also 15 June 1863, p. 101, 15 Feb. 1864, p. 26.

82   *The Nation*, 5 Mar. 1864. See also *The Nation*, 15 Jan. 1876.

83   *The Irish Times*, 5 June 1872.

84   *Ibid.*

85   *Ibid.*

86   An unsuccessful attempt to blow up the statue was made in the year of its unveiling. It was removed from its position in the centre of Leinster Lawn in 1924 to the side adjacent to the Natural History Museum.

87   See *Statement of the proceedings taken to erect and obtain a suitable site for erecting statues thereon, the Gough equestrian statue,* by John Henry Foley Esq, RA [NLI]

88   F. O'Dwyer, *Lost Dublin* (Dublin, 1980), p. 90. Foley had been keen to erect an equestrian statue and wrote to the committee on accepting their offer to execute the statue that, 'I need scarcely repeat how gratifying the task would be to me, and how willing I am to forgo all consideration of profit in my desire to engage myself upon it . . . I feel that the time has arrived for our native country to add to the memorials of her illustrious dead an Equestrian Statue and that Lord Gough at once presents a worthy subject for a memorial.' See the *Statement of the proceedings*, letter dated 25 Aug. 1869.

89   See the *Statement of the proceedings*, letter dated 25 Aug. 1869.

90   *IrB*, 15 Aug. 1879.

91   *Ibid.*

92   *Ibid.*

93   *IrB*, 15 Nov. 1879, p. 347.

94   *The Irish Times*, 23 Feb. 1880. See also the report in *The Irish Times* on 15 Aug. 1986 after the statue was sold by the Office of Public Works to Mr. Robert Guinness for an undisclosed sum believed to be less than £1,000.

95   *The Irish Times*, 6 May 1904, p. 7; *FJ*, 6 May 1904, p. 2.

96   *FJ*, 6 May 1904, p. 2.

97   See *IrB*, 1907, pp. 197 and 291.

98   The names of some of the battlefields are carved around the arch and those of the dead are inscribed in panels over the gates.

99   *The Irish Times*, 17 Feb. 1908, pp. 2–4. The OPW holds a set of files for this monument after 1948 when it was removed from Leinster House. Previous files have been lost. See A. Denson, *John Hughes, sculptor* (Kendal, 1969).

100 *The Times*, 16 Apr. 1900, p. 5.

101 *Ibid.*, p. 11.

102 *Ibid.*, p. 12.

103 *Ibid.*

104 *Weekly Irish Times*, 18 Jan. 1908, p. 19.

105 *Ibid.*

106 *Ibid.* In addition, three small figures representing science, literature, and art were included, reflecting some of the interests of Queen Victoria.

107 *The Irish Times*, 17 Feb. 1908, pp. 2–4.

108 *Ibid.*

109 *Ibid.*

110 *Ibid.*

111 *IrB*, 22 Feb. 1908, pp. 112–13.

112 *Ibid.*

113 J. V. O'Brien, *Dear dirty Dublin. A city in distress, 1899–1916* (London, 1982), p. 35.

114 M. E. Daly, *Dublin: The deposed capital* (Cork, 1985), p. 208.

115 *Ibid.*

Chapter 4 Naming Dublin

1   *IrB*, 15 Aug. 1874, p. 221.

2   *The Irishman*, 12 Dec. 1862.

3   D. Hyde, 'The necessity for de-Anglicising the Irish nation' (1892), quoted in B. Ó Conaire (ed.), *Douglas Hyde, language, lore and lyrics, essays and lectures* (Dublin, 1986), pp. 166–7.

4   Minutes, 1884, no. 375.

5   RPDCD, 1884, vol. III, no. 176.

6   *Ibid.* The report went on to point out that, 'Your committee recommend that the Deputy-Surveyor be directed to report to the Council all cases of streets which, though situated in different parts of the City, bear similar names, with a view to the necessary changes being made to obviate the inconvenience and confusion which arise from such a system of naming streets; also that in future when name-plates of streets are being repainted, he shall be careful to see that in cases of streets named after saints – as in the cases of "St. Kevin Street," "St. John Street," the prefix "St." shall distinctly appear.'

7   Minutes, 1884, no. 391.

8   Minutes, 1884, no. 416.

9   Minutes, 1884, no. 454.

10 Minutes, 1884, no. 470.

11 *Ibid.*

12 Minutes, 1885, no. 23.

13 *Ibid.*

14   Quoted in W.N. Osborough, *Law and the emergence of modern Dublin* (Dublin, 1996), pp. 48–9.

15   The name change took effect on 5 May 1924.

16   Minutes, 1884, no. 229.

17   Minutes, 1885, no. 246.

18   Minutes, 1885, no. 286. The issue of streets with similar names was also addressed by the Corporation, and was the subject of a report which stated that all streets with the prefix 'Saint' were to be distinctly labelled as such, see RPDCD, 1885, no. 40.

19   Minutes, 1885, no. 308.

20   RPDCD, 1885, no. 149.

21   The memorials also pointed to the great inconvenience and annoyance caused to the inhabitants of Upper Temple-street by their letters and parcels being frequently delivered at similar numbers in the lower instead of upper street.

22   Minutes, 1886, no. 107. It is worth noting that the street became known for a time as Chatterton Street [Nov. 1885] after the vice-chancellor who granted the injunction against the renaming of Sackville Street. It remained Chatterton Street until 29 Mar. when the council reversed their decision of 9 Nov. and opted for Hill Street instead.

23   Minutes, 1886, no. 186.

24   RPDCD, 1886, vol. ii, no. 75.

25   Minutes, 1886, no. 186.

26   Minutes, 1888, no. 341.

27   *Ibid.*

28   RPDCD, 1888, vol. iii, no. 110.

29   Minutes, 1888, no. 362.

30   RPDCD, 1901, vol. i, no. 85.

31   Minutes, 1906, no. 92. This name change had been the subject of a report published a year earlier which referred to a memorial that had been received by the committee. It stated that, 'We the undersigned, being a majority in number and value of the ratepayers in the street known as "Montgomery-street," hereby pray the Corporation of the city of Dublin to change its name to Corporation-street. We believe the great improvement effected in the street by the Corporation, together with the change of name, will have the desirable effect of obliterating its evil reputation. We therefore request and pray that it may be called Corporation street,' see RPDCD, 1906, vol. iii, no. 253. The report was carried with an amendment which proposed that the words 'Corporation-Street' be deleted and Foley street substituted.'

32   Minutes, 1911, no. 400. See also RPDCD, vol. iii, 1911, no. 174, Minutes, 1911, no. 760.

33   In 1911 the motion was withdrawn that, 'the name of Great Britain St be changed to Parnell St., and that the names of Rutland Square and Cavendish Row be changed to Parnell Square in recognition of the services to Ireland of the great Irish chief, Charles Stewart Parnell and in accordance with the letter from the Paving committee', Minutes, 1911, no 455. The motion was again postponed on two occasions, Minutes, 1911, no 537, no. 586. See also *The Irish Times*, 3 Oct. 1911, p. 10 and *FJ*, 3 Oct. 1911, p. 2.

34   Minutes, 1911, no. 669. Later in the year protests on behalf of certain residents and traders were received by the council objecting to the proposal to alter the name of that thoroughfare,

Minutes, 1911, no. 717. A memorial was also submitted from residents in Great Britain Street, expressing a desire that the name of that thoroughfare *should* be changed to Parnell Street, 'in recognition of the services to Ireland of the late great leader of the Irish race, C. S. Parnell', Minutes, 1911, no, 719. Further correspondence from those objecting to the name change was received shortly after when a letter was submitted from Mr. W.Q. Fitzgerald, secretary of the Rotunda Hospital, Dublin, embodying a protest from the governess of that institution against the proposed change, Minutes, 1911, no. 718.

35 Minutes, 1900, no. 543.

36 Some weeks later a motion appeared on the order paper in the name of Councillor Nannetti, 'That the committee having charge of the name plates on our streets be instructed to have the names painted in Gaelic in future; and that immediate steps be taken to give effect to this resolution', Minutes, 1900, no. 625. This was later followed by a letter submitted by the Honorary Secretary of the St. Kevin's branch of the Gaelic league, enclosing a copy of a resolution which had been adopted by that body, calling attention to the alleged long delay in affixing the names of the streets and thoroughfares in Irish characters. It was moved that the letter be referred to the paving committee, Minutes, 1900, no. 634.

37 RPDCD, 1900, vol. III, no. 85.

38 Minutes, 1901, no. 23.

39 Minutes, 1904, no. 126.

40 Minutes, 1904, no. 126.

41 Minutes, 1909, no, 281.

42 Minutes, 1909, no. 424.

Chapter 5 Symbolising the state

1 *FJ*, 26 Apr.–5 May 1916.

2 Damage was estimated at £2 million and the centre of the city was left in ruins. In Dublin, the Civil War caused damage estimated at just over £5 million. Sixty-four people were killed and 500 wounded, see J. J. Lee, *Ireland 1912–1985* (Cambridge, 1989).

3 Fighting in the Civil War continued throughout the country during 1923 resulting in the deaths of many of the leading figures on both sides.

4 Much of the destruction was confined to Lower Sackville Street, the city's central thorough-fare. From Bachelor's Walk up to and including the General Post Office was reduced to ruins. For a summary of the course of the rebellion see Department of External Affairs, *Cuimhneachán 1916, Commemoration: a record of Ireland's commemoration of the 1916 Rising* (Dublin, 1966).

5 *OPW annual report*, 1916–17, p. 12 (hereafter *OPW*).

6 The rebuilding of Sackville Street was the subject of much debate, in particular over the issue of whether the street should follow a uniform plan or whether individual architects should be allowed free reign. A compromise was eventually reached under the direction of the city architect, Horace O'Rourke, in the aftermath of the War of Independence. See F. McDonald, 'From the rubble of the rising', *The Irish Times*, 27 Apr.1991; M. Shaffrey, 'Sackville Street/O'Connell Street', *Irish arts review yearbook*, 1988, pp. 144–56; M. Bannon (ed.), *The emergence of Irish planning 1880–1920* (Dublin, 1985); *Planning: The Irish experience,*

1920–1988 (Dublin, 1989); 'The rebuilding of central Dublin' letter from the Council of the Royal Institute of Architects of Ireland, *RIBA Journal* 10 June, 23 (1916): 262; W. A. Scott, 'Reconstruction of O'Connell Street Dublin: a note', *Studies* Nov. (1916): 165; R. M. Butler, 'The reconstruction of O'Connell Street, Dublin', *Studies* Dec. (1916): 570–6.

7   *The Irish Times*, 26 May 1921, pp. 3–7.

8   *Ibid.* At the time the Custom House housed the Department of Inland Revenue, Local Government Board, Assay Office, Estate Duty Office, Income Tax Office and the Stationery Office. Its destruction was ordered by Dáil Éireann as 'a military necessity' and they regarded it as 'one of the seats of an alien tyranny'.

9   *IrB*, 4 June 1921, pp. 1, 393–7; 18 June 1921, pp. 417–18, 421; 2 July 1921, pp. 449; 28 Jan. 1922, pp. 46, 61; 25 Feb. 1922, p. 113; 20 May 1922, p. 1; *IrB*, 7 Oct. 1922, p. 670; *The Irish Times*, 26 May 1921, pp. 3–7.

10   The controversial treaty came into effect in December 1921 formally creating the Irish Free State on 6 Dec. 'The Articles of Agreement for a treaty between Great Britain and Ireland', as it was formally known, led to the political split and eventually the civil war. The war saw 600 lose their lives, 3,000 injured and damage estimated at £30 millions.

11   Commentators have speculated subsequently that, given the building contained priceless records of huge symbolic significance, the irregulars must have felt that they could safely occupy this building.

12   See *IrB* 15 July 1922, pp. 493–7, p. 493. See also reports in *IrB*, 15 July 1922, pp. 481–2, 489; 15 July 1922, pp. 294–7; *IrB*, 29 July 1922, pp. 513–14, 517–18; *IrB*, 16 Dec. 1922, p. 857 and *The Irish Times*, 4 July 1922. Substantial quantities of irreplaceable public records were lost in the bombardment of the building.

13   *IrB*, 15 July 1922, p. 493. See also 'The Sackville Street holocaust', in the *IrB*, 15 July 1922, p. 490. The Annual report of the OPW for the years 1922–3 noted the destruction which affected the city: 'This destruction began as part of the movement against the Free State government . . . In Dublin we have endeavoured, we believe with some success, to provide temporary accommodation for the officials formerly housed in the GPO, the Custom House and the Four Courts pending the reconstruction of those important buildings, see *OPW*, 1923, p. 5.

14   *IrB*, 15 July 1922, p. 481.

15   See T. Brown, *Ireland: a social and cultural history* (London, 1985). On visual representations of identity after 1922 see B. P. Kennedy, 'The Irish Free State 1922–1949: A visual perspective', in B. P. Kennedy and R. Gillespie (eds), *Ireland: art into history* (Dublin, 1994), pp. 132–54.

16   Kennedy, 'Irish Free State', pp. 132–54.

17   *FJ*, 12 Dec. 1922 editorial comment.

18   *Greater Dublin reconstruction scheme described and illustrated*, hereafter GDRM (Dublin, 1922), p. 10.

19   See *IrB*, 26 Aug. 1922, p. 582 and 30 Dec. 1922, pp. 891–2.

20   L. J. Vale, *Architecture, power and national identity* (London, 1992), p. 10.

21   Before the Treaty of 6 Dec. 1921, Dáil Éireann had met in a number of locations. Michael Collins, then Minister for Finance, decided that the most convenient place for the Provisional Parliament would be the lecture theatre of the Royal Dublin Society at Leinster House.

Proceedings opened there on 9 Sept. 1922, see *IrB*, 13 Jan. 1923, p. 1; 28 June 1916, p. 581; 24 July 1916, p. 539.

22 GDRM, pp. 5–6.

23 See *OPW*, 1922–3, p. 6.

24 The Royal College of Science on Merrion Street had been completed in March 1922. See *OPW*, 1922, p. 8, which describes it as 'one of the most important building schemes ever undertaken in Dublin in the past century'. See also *OPW*, 1922, pp. 9–10, pp. 16–17. It was requisitioned for governmental use for most of the 1922–3 academic year .

25 *OPW*, 1924, pp. 3–4. 'Oireachtas' refers to the two legislating bodies of the state, the Dáil (lower house of Parliament) and the Seanad or Senate (upper house of Parliament).

26 Dublin Corporation's City Archives hold the records of this group and many of the newspaper cuttings and documents published about it, see City Archives, GRC/1, 2, 3, 4, 5.

27 GDRM, pp. 5–6.

28 *Ibid.*, p. 7.

29 *Ibid.*

30 *Ibid.*

31 *Ibid.*

32 The Civic Survey was eventually carried out and published in 1925. An extensive survey, it presented the fundamental facts concerning the City and Environs of Dublin and was intended as a compendium of information on all aspects of the city's life and as the prelude to a comprehensive town plan, keeping in line with the Geddessian principle of survey, analysis and plan. See M. Bannon, (ed.), *Planning: the Irish experience 1920–1988* (Dublin, 1989), p. 21.

33 The Castle had been ceremoniously handed over to the Irish government by the British on 16 Jan. 1922.

34 *Dáil Debates*, 27 Mar. 1923, cols. 2559–62.

35 *Dáil Debates*, col. 2560.

36 *Dáil Debates*, col. 2560.

37 *IrB*, 13 Jan. 1923, p. 1.

38 Cited in *IrB*, 24 July 1923, p. 530. The *IrB* also noted that: 'We are delighted to see that public opinion is steadily rallying in favour of the "Old House" and we trust that the result of the deliberations of the Committee appointed by the government will be an unanimous recommendation in favour of College Green', p. 583.

39 *Dáil Debates*, 10 July 1923, col. 415–16.

40 *Dáil Debates*, 27 July 1923, cols 1549–71., col. 1550.

41 *Dáil Debates*, 8 Aug. 1923, cols 1993–4. See also *IrB*, 11 Aug. 1923, p. 1 and 25 Aug. 1923, p. 630.

42 *IrB*, 29 Dec. 1923, p. 1005.

43 *Dáil Debates*, 24 Jan. 1924, cols 564–99, col. 565.

44 *Dáil Debates*, 24 Jan. 1924, cols 571–2.

45 *Dáil Debates*, 24 Jan. 1924, cols 574–6.

46 *Ibid.*

47 *Dáil Debates*, 19 Mar. 1924, cols 2139–55.

48 *Dáil Debates*, 19 Mar. 1924, cols 2142–4.

49   *Dáil Debates*, 19 Mar. 1924, col. 2902.

50   *Dáil Debates*, 19 Mar. 1924, col. 2902.

51   *IrB*, 8 Apr. 1924, p. 1, See also 19 Apr. 1924.

52   *Dáil Debates*, 25 June 1924, cols 2941–50; 10 July 1924, cols. 911–35. See also *OPW*, 1925, pp. 3–4; 10 July 1924, cols 911–35; and see also *IrB*, 23 June 1924, p. 558.

53   P. Abercrombie, A. Kelly and S. Kelly, *Dublin of the Future* (Dublin, 1922), p. vi.

54   *Ibid.*, p.3.

55   See M. J. Bannon, 'The genesis of modern Irish planning', in M. J. Bannon (ed.), *A hundred years of Irish planning* (Dublin, 1985), pp. 189–262.

56   Abercrombie et al., *Dublin of the Future*, p. ix.

57   *Ibid.*, p. x.

58   *Ibid.*, p. 36.

59   *Ibid.*

60   *Ibid.*

61   *Ibid.*, p. 38.

62   The *IrB* noted in 1922 that the late Archbishop, Dr Walsh, was actually about to acquire the Ormond Market site from the Corporation, and was only deterred by the difficulty he experienced in securing an open frontage to the Quays, 29 July 1922, p. 513.

63   *IrB*, 29 July 1922, p. 513. *Ibid.* This was rejected in some quarters, most notably in the *IrB*, which saw it as a last resort site that could only be utilised if no better site was found and argued that it was not central enough and that Upper Sackville Street would be an ideal site for the cathedral, see also p. 510.

64   *IrB*, 15 Jul. 1922, p. 498.

65   *Ibid.* See also 9 Sept. 1922, cover page; 4 Nov. 1922, p. 754; 30 Dec. 1922, pp. 881–2. See also 4 Nov. 1922, p. 754.

66   *IrB*, 30 Dec. 1922, p. 881.

67   *The Irish Times*, 13 July 1930.

68   *IrB*, 19 July 1930, p. 644. A number of sites had been considered by this point, among them Mountjoy Square and the Rotunda Gardens.

69   See *The Irish Times*, 6 Apr. 1974. The issue had previously re-emerged in the context of P. Abercrombie et al., *Dublin sketch development plan* (Dublin, 1941).

70   *IrB*, 26 Aug. 1922, p. 382.

71   *IrB*, 7 Oct. 1922, front page. See 'The Custom House, Dublin', p. 670, where it is noted that 'Dublin without its Custom House! It would be better to hammer blows upon the Cupola of St. Peter's.'

72   *IrB*, 'The Dáil and public works', 6 Dec. 1922, p. 851.

73   Attempts to reconstruct the GPO had begun in 1920. See *OPW*, 1920–1, p. 11.

74   See *IrB*, 20 Dec. 1919, p. 570; also for reports on progress made in the rebuilding work, 4 June 1921, p. 386; 11 Feb. 1922, front page; 25 Feb. 1922, p. 116; 25 Mar. 1922, p. 185–6 and 8 Apr. 1922, p. 250.

75   'The rebuilding of the Dublin Post Office', *IrB*, 15 Nov. 1924, pp. 973–4, p. 973. See also, 'Reconstruction. No. 2 – The General Post Office, 4 Apr. 1925, p. 262; 11 July 1925, front page. Work began on the front block in Aug. 1925, followed by Henry Street some months later. In

1927 'out of the ashes of the fire in 1916', the General Post Office was gradually recreated, see *IrB*, 25 June 1927, p. 478; *OPW*, 1926–7, p. 8. For a further report on the reconstruction, see *OPW*, 1928, p. 9. The shopping arcade was carried out between 1928–9. Further reconstruction work was carried out in 1930, see *OPW*, 1931–2, p. 11, 1932–3, p. 8.

76   *OPW*, 1933–4, pp. 7–8.

77   *IrB*, 24 Mar. 1925, p. 222; see also *IrB*, 25 July 1925, front page; 17 Aug. 1929, p. 732.

78   *OPW*, 1925–6, p. 6.

79   *Ibid.*, appendix A, pp. 20–1.

80   *OPW*, 1926–7, pp. 5–6, see also 1927–8, pp. 4–5.

81   *OPW*, 1928–9, pp. 6–7. This report gives a detailed description of the origins of the building, while appendix AA gives an account of the reconstruction programme carried out under the supervision of T. J. Byrne, Principal architect of the OPW, see pp. 36–9. It noted that it was decided to use Ardbraccan rather than Portland stone in the reconstruction of the dome while new carvings of 'faithful replicas of Edward Smyth's renowned work were executed by Dublin sculptors. The four dialed clock is electrically synchronised with Dublin mean time and provision for night illumination has been made', p. 39. See also *OPW*, 1930–1, p. 27.

82   *IrB*, 7 Mar. 1925, p. 169.

83   Some saw the Four Courts as a relic of the British empire with no place in the newly established Irish Free State; others debated whether a Catholic Cathedral ought to be erected, see *IrB*, 29 July 1922, p. 510.

84   *The Builder* (London), 7 July 1922.

85   *IrB*, 15 July 1922, p. 498.

86   For a discussion of the rebuilding and the architectural implications, see H. Cotter, 'The reconstruction of the Four Courts following its destruction in 1922', unpublished BArch Dissertation, School of Architecture, UCD, 1989.

87   Byrne was the principal architect of the OPW, and had responsibility for overseeing the reconstruction of the Custom House and General Post Office.

88   Lecture given by T.J. Byrne to the Institute of Civil Engineers of Ireland, 1928, p. 10, reprinted in *IrB*, 21 Jan. 1928, pp. 54–60. See also 'Reconstruction: No. 1 – The Four Courts', *IrB*, 7 Mar. 1925, p. 169 and 'Reconstruction of the Four Courts, Dublin', *IrB*, 25 July 1925, p. 617. See also *OPW*, 1931–2, p. 7. The flanking blocks to the west and east were started first and as had occurred with the Custom House many of the internal walls were removed and replaced by a steel framework with lightweight partitions. The OPW wrote in their report for 1926, 'The plan adheres as closely as possible to the central conception of Gandon's original design, viz., the Four Courts grouped round a circular central hall, which is surmounted by a dome, but adds a Supreme Court in an axial position and two other courts . . . The court offices and other accessory rooms will be greatly improved in spaciousness and simplicity of arrangement, as compared with those of the old four courts building, by the adoption of modern methods, particularly steel frame construction inside the blocks. The floors will all be of concrete, which will make the building more resistant to fire', *OPW*, 1925–6, p. 9.

89   *OPW*, 1931–2, pp. 7–8.

90   See *OPW*, 1927–8, p. 5 and 1928–9, p. 8 and 1929–30, p. 6.

91   *IrB*, 24 Oct. 1931, p. 917.

92   Darrell Figgis in an address to the AAI, 28 Mar. 1922, quoted in S. Rothery, *Ireland and the new architecture, 1900–1940* (Dublin, 1991), p. 1.

93   S. O'Laoire, 'Building on the edge of Europe', in J. Graby (ed.), *Building on the edge of Europe/Construire à la frange de l'Europe': a survey of contemporary architecture in Ireland embracing history, town and country* (Cork, 1996), pp. 118–67, p. 145.

94   H. Campbell, 'Irish identity and the architecture of the new state', in A. Becker, J. Olley and W. Wang (eds), *Twentieth century architecture: Ireland* (Munich, 1997), pp. 82–5, p. 83. It is important to note, however, that the Free State government did forge ahead with comprehensive programmes of school and hospital building. See AAI, *Public works, the architecture of the Office of Public Works, 1831–1937* (Dublin, 1987); Government of Ireland *Building for Government* (Dublin, 1999); J. O'Sheehan and E. de Barra, *Oispidéil na hÉireann: Ireland's hospitals, 1930–1955* (Dublin, 1956). It is notable also that the state, and particularly Dublin Corporation, embarked on a massive programme of housing provision which, significantly, made use of an international architectural style; see S. Rothery, *Ireland and the new architecture, 1900–1940* (Dublin, 1991), pp. 140–54.

95   S. Walker, 'Architecture in Ireland, 1940–1975', in A. Becker et al., *Twentieth century architecture*, p. 23.

96   *OPW*, 1935–6, pp. 7–8. See also *IrB*, 24 Aug. 1936, p. 737. The department's accommodation was spread over a number of buildings and in 1929 it had become clear that alternative, purpose built accommodation was required. The OPW hold the file regarding the origins and evolution of the project, see A15.1/1929–42. See also A. Rolfe and R. Ryan, *The Department of Industry and Commerce, Kildare Street, Dublin* (Dublin, 1992).

97   *IrB*, 11 Jan. 1936, p. 7.

98   *Ibid.*

99   The OPW, as Rolfe suggests, was undoubtedly aware of the embarrassment associated with what had happened during the competition for the design of the National Museum in the 1880s when the architects of the five designs selected from 67 submissions all turned out to be unknown Englishmen. The architects were eventually paid a total of £1,450 and a second competition was held which was won by the Cork architects, T. N. Deane and son. See Rolfe and Ryan, *Department of Industry and Commerce*, p. 13.

100   J. R. Boyd-Barrett had an extensive practice in Cork; second prize was awarded to Alfred E. Jones and Stephen S. Kelly and third to Ms E. Grace Butler. All the designs were exhibited at the Metropolitan School of Art for a week. This exhibition was reviewed and commented on fully by 'Wisbech' in the *Irish Builder and Engineer*, see *IrB*, 25 Jan. 1936, pp. 62, 65, 67–8 and 8 Feb. 1936, p. 102.

101   *IrB*, 11 Jan. 1936, p. 7. See also supplement to the *IrB*, 25 Jan. 1936 for illustrations of the plans for the building and 21 Nov. 1942 for a progress report.

102   Although it had been seen as the architectural style of fascism, it was the style then prevailing in Europe and the English-speaking world and does not necessarily have any political overtones. Rothery writes of it as a surprising choice, 'given the clearly modern image actively promoted by all of the other building types of the new state; the hospitals, schools, and of course the new airport', S. Rothery, *Ireland and the new architecture, 1900–1940* (Dublin, 1991), p. 89.

103   Tenders for construction work were invited on 7 Feb. 1938.

104   These were Laurence Campbell, Albert Power, Oliver Sheppard, George Smith and Séamus Murphy. In the wake of Sheppard's death, Gabriel Hayes was asked to take part.

105   R. Ryan, 'Architectural appraisal', in Rolfe and Ryan, *Department of Industry and Commerce*, p. 37.

106   S. O'Laoire, 'Building on the edge of Europe', in J. Graby (ed.), *Building on the edge of Europe/Construire à la frange de l'Europe: a survey of contemporary architecture in Ireland embracing history, town and country* (Cork, 1996), pp. 118–67, p. 151.

107   Ireland's last appearance at a world fair had been at Chicago in 1882. For accounts of the 1939 pavilion see N. Sheaff, 'The Shamrock building', *Irish Arts Review* 1 (1984): 26–9; Rothery, *Ireland and the new architecture*, pp. 220–8.

108   Up to October 1937 no architect had been appointed to the task and it was in turn suggested that an architectural competition be held. During that month a meeting was held at which it was recommended that an Irish architect be given the commission and the OPW was requested to submit a list of possible candidates. By April 1938 no decision had been made and the issue was seen as urgent, as the pavilion would not be completed in time for the opening of the fair. In May of 1938 Scott was appointed. His detailed plans of the Irish design are preserved at the Irish Architectural archive.

109   Nicholas Sheaff interview, quoted in D. Walker, *Michael Scott architect* (Kinsale, 1995).

110   Walker, *Michael Scott*, p. 96

111   Scott commissioned Eric Gill, a celebrated English sculptor and typographer to carry out the typography.

112   *The Irish Press*, 13 May 1939.

113   P. Abercrombie, 'The Dublin town plan', *Studies* 31 (1942): 155–60, p. 155.

114   Abercrombie et al., *Dublin sketch development plan*.

115   The planners were concerned with the unlimited nature of growth in the city. They proposed the imposition of a green belt between four and six miles wider circling the built-up area and urged the introduction of satellite towns to cope with future population growth.

116   Abercrombie, 'Dublin town plan', p. 160.

117   Abercrombie et al. *Dublin sketch development plan*, p. 36.

118   *Ibid.*

119   *Ibid.*, p. 37.

120   Abercrombie, 'Dublin town plan', p. 158.

121   Abercrombie et al. *Dublin sketch development plan*, p. 37.

122   Abercrombie, 'Dublin town plan', p.159.

123   *Ibid.*, p. 160.

124   *Ibid.*

125   Abercrombie et al., *Dublin sketch development plan*, p. 36.

126   *Ibid.*, p. 38. In an appendix to the town planning report this proposal was removed from the proposed statutory plan, see p. 62.

127   Abercrombie et al., *Dublin sketch development plan*, p. 38. Some years later in 1955 the architects Jones and Kelly prepared a design for civic offices located further west from

Abercrombie's proposal on a site on Wood Quay. It was not until 1969, however, that a competition was held. Stephenson Gibney won it and their controversial plan was eventually completed in the 1980s.

128   *IrB*, 21 Dec. 1940, p. 759.

129   L. F. Giron, 'Comments on the foregoing article, No. 2', *Studies* (1942): 163–4, p. 163.

130   J. J. Robinson, 'Comments on the foregoing article, No. 3', *Studies* (1942): 164–6, p. 164.

131   *Ibid.*, p. 165.

132   *Ibid.*

133   *Ibid.*

134   *Ibid.* p. 166.

135   C.P. Curran, 'Comments on the foregoing article, No. 4', *Studies* (1942): 167–70, p. 169.

136   *Ibid.*, p. 170.

137   R. McGrath, 'Dublin panorama: an architectural review', *The Bell* 2, 5 (1941): 33–48, p. 35.

138   The election of 1948 saw the Fianna Fáil administration removed from power after sixteen years. A coalition government made up of Fine Gael, Clann na Poblachta, Labour, National Labour, Clann na Talmhan and some independents, under the leadership of John A. Costello, came to power in its place. A year earlier, while on a state visit to Canada, the President, Sean T. O'Kelly announced in Ottawa that it was his intention to repeal the External Relations Act and with it secede from the Commonwealth. The External Relations Act which had been passed by the de Valera led administration in 1938, had 'created a subtle, if not ambiguous, relationship with Britain and the Commonwealth', J. A. Murphy, *Ireland in the twentieth century* (Dublin, 1989), p. 124.

139   See T. Brown, *Ireland: A social and cultural history 1922–1985* (London, 1985), p. 212.

140   See F. S. L. Lyons, 'The years of readjustment', in K. B. Nowlan and T. D. Williams (eds), *Ireland in the war years and after, 1939–51* (Dublin, 1969), pp. 67–79; J. J. Lee (ed.), *Ireland 1945–70* (Dublin, 1979).

141   Whitaker declared in 1959 that 'After thirty-five years of active government people are asking whether we can achieve an acceptable degree of economic progress. The common talk among parents in the towns, as well as in rural Ireland, is of their children having to emigrate as soon as their education is completed in order to secure a reasonable standard of living.' This ushered in an era of large-scale state involvement in the planning and financing of economic development. See B. M. Walsh, 'Economic growth and development 1945–1970', in J. J. Lee (ed.), *Ireland 1945–1970* (Dublin 1979), p. 29.

142   Walker, 'Architecture in Ireland', p. 23.

143   RIAI, *Year-book*, 1946, editorial by Raymond McGrath.

144   Walker, 'Architecture in Ireland', p. 23.

Chapter 6 Scripting national memory

1   This was the location where 14 out of 16 leaders of the rebellion had been buried in quick lime shortly after the rebellion.

2   Sean T. Lemass, 'Foreword', in Department of External Affairs, *Cuimhneachán 1916–1966: a record of Ireland's commemoration of the 1916 Rising* (Dublin, 1966), p. 11.

3   It was eventually erected on St Stephen's Green in 1967 opposite the birthplace of Emmet and in the same year as a monument dedicated to Wolfe Tone was unveiled in the capital in the north east corner of St Stephen's Green. See M. Allen, 'Jerome Connor – one', *The Capuchin Annual* (1963): 347–68; 'Jerome Connor – two', *The Capuchin Annual*, (1964): 353–69; 'Jerome Connor – three', *The Capuchin Annual* (1965): 365–87.

4   *The Irish Times*, 14 Aug. 1923, p. 5.

5   See OPW files: Cenotaph, Leinster Lawn, A/25/22/2, vol. 1.

6   The Cumann na nGaedheal party in power at the time had instigated a Gaelicisation programme. In Nov. 1924 the declaration was issued that, 'The Organisation and the government are pledged to co-ordinate, democratize and gaelicise our education. It is now possible for the child of the poorest parents to pass from one end of the educational ladder to the other, and the Irish language has been restored to its own place in Irish education.' See T. Brown, *Ireland: a social and cultural history 1922–1985* (London, 1985), pp. 48–56.

7   The Celtic cross was perhaps the most public manifestation of a government policy to honour the heroes of the years of civic unrest. See S. Bhreathnach-Lynch, 'Executed: the political commissions of Albert G. Power', in *Éire–Ireland* (1994): 44–60 and 'The art of Albert Power, 1881–1945: A sculptural legacy of Irish Ireland', in B. P. Kennedy and R. Gillespie, *Ireland: art into history* (Dublin, 1996), pp. 118–32.

8   Ten days after the unveiling of the monument, a wreath-laying ceremony was held to commemorate the first anniversary of Collins's death. Following the assassination of Kevin O'Higgins some years later in 1927, a third plaque was fixed to the plinth.

9   *The Irish Times*, 3 Aug. 1923, p. 4.

10   *Ibid.*

11   *The Irish Times*, 14 Aug. 1923, pp. 4–5.

12   *Ibid.*

13   Collins and Griffith were on opposite sides to de Valera in the Civil War of 1922–3. The Fianna Fáil administration was inevitably going to have different priorities to the Cumman na Gaedheal administration which had been on the pro-Treaty side in the war. See OPW files, Cenotaph on Leinster Lawn, A/25/22/2, vol. 1. ref. No. 1, 'The Griffith–Collins–O'Higgins Memorial'.

14   In 1933 the government authorised the opening of the lawn from 12 noon to 4 p.m. on Sunday, 13 Aug., to a limited number of people who applied to the OPW for permission. A dispute arose, however, when the Director of the National Guard signalled his intention to organise a parade and wreath laying ceremony. The parade was banned by the government and subsequently cancelled by the National Guard. The OPW holds no record of any subsequent ceremonies at the cenotaph and when the new obelisk was erected in 1950 there was no unveiling ceremony. See OPW, Cenotaph.

15   The Civil War had created a divide in Irish politics that had not healed by the middle of the century. This memorial sought to honour two of the chief victims of the war who had been fierce opponents of de Valera. It fell to the de Valera administration to put in place the plans for the new memorial.

16   See *Dáil Debates*, 1944, col. 1419; 1945, col. 165; 1947, cols 774–5, 366, also OPW, Cenotaph.

17   This work was carried out by the sculptor Laurence Campbell.

18 A. Martin (ed.), *W.B. Yeats, Collected Poems* (London, 1989).

19 *The Irish Times*, 22 Apr. 1935, p. 6.

20 J. Turpin, 'Cuchulainn lives on', in *Circa*, 1994, 69, pp. 25–9, pp. 26–7.

21 *Ibid.*, p. 28.

22 *United Ireland Journal*, 20 Apr. 1935, pp. 1–2. See also S. Bhreathnach-Lynch, 'Commemorating the hero in newly Independent Ireland: expressions of nationhood in bronze and stone', in L. McBride (ed.), *Images, icons and the Irish nationalist imagination* (Dublin, 1999), pp. 148–65.

23 *The Irish Times*, 22 Apr. 1935, p. 7.

24 *Ibid.*, p. 7.

25 *Ibid.*, p. 8.

26 *Ibid.*

27 *Ibid.*

28 *Ibid.*, p. 7.

29 *Ibid.*

30 *The Irish Press*, 18 Apr. 1935.

31 S. Bhreathnach-Lynch, 'Executed: the political commissions of Albert G. Power', in *Éire–Ireland* (1994): 44–60, p. 53.

32 The monument was damaged in 1945, see *Dáil Debates*, 1945, cols 2241–2. It was subsequently replaced in 1956 with a similar memorial by Seamus Murphy.

33 *The Irish Times*, 4 July 1932, p. 3.

34 A number of the city's main train stations were renamed after the leaders of the 1916 rising, for example, Kingsbridge became Heuston, Amiens Street station became Connolly, and Westland Row became Pearse Street.

35 *Irish Independent*, 14 Mar. 1953.

36 Speech of President Sean T. O'Kelly, reported in *The Irish Times*, 21 May 1956, p. 7.

37 *Dáil Debates*, 1954, cols 1289–90.

38 *Ibid.*

39 That the urban landscape was used as a means of preserving a commemorative record of the 1916 Rising was illustrated not only in these prominent examples but also in less conspicuous monuments. For example, a bronze memorial by Werner Schreremann was unveiled in 1960 on Parnell Square, East, commemorating the foundation of Oglaigh na hÉireann (the Irish Volunteers), which took place close by in the Rotunda Gardens in 1913.

40 D. MacDonagh, 'Young Ireland' from *The Irish Times*, 7 Sept. 1945.

41 The statue dedicated to George II had been destroyed in 1937.

42 Thomas Davis was the editor of *The Nation* newspaper and co-founded the Young Ireland movement, a group of young nationalists in the mid-nineteenth century. See *Pictorial record of the centenary of Thomas Davis and Young Ireland* (Dublin, 1945).

43 *The Irish Times*, 10 Sept. 1945, p. 1.

44 *Ibid.*

45 *The Irish Times*, 13 Sept. 1945, p. 1.

46 For a record of the parliamentary debate about the monument see *Dáil Debates*, 1960, cols 948–9, no. 12. See also 1962, col. 475, no. 24; col. 20, no. 14; 1963, cols 1284–5, no. 27; cols

19–20, no. 28; cols 581– 2, no. 8. See OPW files: Thomas Davis monument: A/99/3/5, vols I and II, A/99/3/31, vols I and II.

47   The bronze figure of Davis, the fountain reliefs and the figures of the heralds were modelled by the sculptor Edward Delaney. The figure of Davis was cast by the Fonderia d'Arte in Milan. The sculptor himself cast the heralds and the bronze reliefs. The architectural treatment was carried out by the Works section of the OPW in collaboration with the sculptor and the inscription on the pedestal was by Michael Biggs.

48   Department of External Affairs, *Cuimhneachán*, p. 62.

49   *Ibid.*

50   O'Donovan Rossa was born in Rosscarberry, Co. Cork in 1831. In 1856 he founded the Phoenix Society, a literary and political group which was subsumed by the republican movement in the mid-nineteenth century.

51   See *IrB*, 15 Aug. 1953, p. 839. See also OPW, A/25/22/2/48, A/ 25/22/2/3, A/25/22/48.

52   *The Irish Times*, 7 June 1954, p. 7.

53   *Ibid.*

54   Department of External Affairs, *Cuimhneachán*, p. 11.

55   *Ibid.*

56   *IrB*, 7 Sept. 1946, p. 558.

57   *Ibid.*

58   *Ibid.*

59   *Ibid.*

60   *Ibid.*

61   *Dáil Debates*, 1946, cols 425–6.

62   *Dáil Debates*, 1949, cols 2054–5.

63   *Ibid.*

64   *Ibid.*

65   *Ibid.*

66   *Dáil Debates*, 1949, cols 181– 2.

67   *The Irish Times*, 12 Apr. 1966, p. 1.

68   Department of External Affairs, *Cuimhneachán*, p. 46.

69   Report on the sculpture for the proposed garden of Remembrance, in OPW files: Garden of Remembrance.

70   *Ibid.*

71   *Ibid.*

72   *Ibid.*

73   *Ibid.*

74   *Ibid.*

75   *The Irish Times*, 12 July 1971, p. 11.

76   *Ibid.*

77   The OPW holds a substantial number of files covering all aspects of the evolution and subsequent maintenance of the war memorial. See also K. Jeffery, 'Commemoration: turning the 11 Nov. into the 12th July', in K. Jeffery, *Ireland and the Great War* (Cambridge, 2000), pp. 107–43.

78   See M. Heffernan, 'For ever England: the Western Front and the politics of remembrance in Britain', *Ecumene* 2, 3 (1995): 293–323.

79   See Lt Col. Boydell, *The Irish national war memorial: its meaning and purpose. A permanent reminder of the splendor of Irish valour and a visible expression of our unparalleled national unity*, NLI and Irish Architectural Archive file, p. 17.

80   *Ibid.*, p. 23.

81   *Ibid.*, p. 19.

82   The progress of the scheme can be charted in *IrB* throughout the 1920s and 1930s. In 1921 the journal noted that 'the proposal had been made some time ago to erect a memorial to the memory of the men of the city and county who had fallen in the Great War. Money was collected and the proposal then made that the memorial should take the form of a Soldier's Home', *IrB*, 1 Jan. 1921, p. 6. The journal also commented upon the fact that, 'The Great War has resulted in a vast crop of monuments, good, bad and indifferent, the vast majority exceedingly bad. Only a few stand out as possessed of any degree of merit, either as regards originality or academic correctness. The vast majority are commonplace to the last degree . . . in Ireland we have comparatively few war memorials. Any we have seen are decidedly bad – very bad indeed', *IrB*, 16 July 1921. The suggestion was also made that an Irish-made memorial be erected at Ginchy where so many men of the Irish divisions fell, see *IrB*, 25 Aug. 1923, p. 1. See also *IrB*, 25 July 1925, p. 594 and *IrB*, 31 Oct. 1925.

83   *The Irish Times*, 30 Mar. 1927 and *IrB*, 2 Apr. 1927, p. 222.

84   The *IrB* stated that 'It now remains for the promoters to start afresh upon a worthy scheme that will command the unanimous support of all, or nearly all, the subscribers, ex-servicemen and others directly interested', 2 Apr. 1927, p. 222. Two years later it was proposed that a housing scheme for the benefit of ex-service men might be an appropriate war memorial, *IrB*, 9 Nov. 1929, p. 1. While a practical solution to the main problem was pending it was found possible to carry out other memorial work using interest accruing from the invested Trustee funds. Memorials dedicated to the Irish officers and soldiers who fell in the war were erected on the battlefields 'through the active instrumentality of the gallant and distinguished commander of the 16th (Irish) Division, Major General Sir William B. Hickie, K.C.B'. In addition, a complete record in eight volumes of the names and services of every Irish officer and soldier who fell in the war was collected, published and circulated by the Committee. These volumes were accepted by the authorities of a hundred different centres and sets of the volumes are preserved at the British Museum, the Vatican, Washington, Paris, Brussels and 'in our Irish cities and in other suitable places. This work involved much labour and expense but it met with public approval and commendation', see Boydell, *Irish national war memorial*, p. 23.

85   *IrB*, 13 Sept. 1930, pp. 793–4. See also *IrB*, 19 Dec. 1931, p. 1096d. Regret was expressed that an Irish architect was not retained for the scheme, see *IrB*, 2 Jan. 1932, p. 1.

86   De Valera's government accepted the project in its entirety despite objections from extreme republicans. In 1932 the Lord Mayor ruled out of order a letter appearing on the notice paper from the Hon. Sec. of Clann na Gaedhael, forwarding on behalf of that organisation a protest against the erection in Dublin of the Great War memorial, Minutes, 1932, no. 26.

87   Boydell, *Irish national war memorial*, p. 31.

88   The war stone is approached from a temple added in the 1980s, which contains a quotation form the work of the war poet, Rupert Brooke.

89   An essential condition of the scheme was that for a certain number of days each year the memorial plot would be at the disposal of ex-Servicemen of the Great War for their annual gatherings to commemorate fallen comrades.

90   Boydell, *Irish national war memorial*, p. 33.

91   *Ibid.*

92   See *IrB*, 4 Feb. 1939, p. 87; 15 May 1937, p. 446.

93   See 'Dublin's forgotten park', *Sunday Tribune*, 14 Apr. 1985, p. 9; also *Sunday Press*, 8 Aug. 1985; 'Dublin's secret garden', *The Irish Times*, 27 Mar. 1985; *The Irish Times*, 22 Feb. 1992, *Evening Herald*, 15 May 1986 and various letters to the editor of *The Irish Times* in May 1985 on the neglect of the war memorial garden.

94   See *The Irish Times*, 28 Apr. 1995.

95   S. Bhreathnach Lynch, 'Commemorating the hero in newly Independent Ireland: expressions of nationhood in bronze and stone', in L. McBride (ed.), *Images, icons and the Irish nationalist imagination* (Dublin, 1999), pp. 148–65, p. 148.

96   D. Fitzpatrick, 'Commemoration in the Irish Free State: a chronicle of embarrassment', in I. McBride (ed.), *History and memory in Modern Ireland* (Cambridge, 2001), pp. 184–204, p. 186.

Chapter 7   Removing icons of empire

1   *Evening Herald*, 12 Dec. 1931.

2   *The Irish Times*, 1923, cited in R. Cathcart, *William III and Ireland* (Dublin, 1990).

3   See *The Irish Times*, 12 Nov. 1928, and various references in 'The Irishman's Diary' section of the paper on 14, 15 and 16 Nov. 1928. Not only were the statues of George II and William III the target of attack, but also 'a third explosion took place in Herbert Park, Ballsbridge, where an ornamental fountain, erected to commemorate the visit of King Edward VII to the international exhibition in 1907 was blown up', see *The Irish Times*, 12 Nov. 1928, p. 9.

4   *Ibid.*

5   *The Irish Times*, 14 Nov. 1928, see also 15 Nov.

6   *Irish Independent*, 5 Jan. 1929, p. 9. A photograph in *The Irish Times* shows 'the statue of King William of Orange, in College Green, Dublin, which was damaged by an explosion on Armistice Day, "in splints" ready for removal', see 30 Nov. 1928. The accompanying report stated that, 'King William III . . . is about to disappear from its present site . . . it is the intention to remove the statue to the Corporation yard in Hanover Street. The ultimate fate of the statue cannot at the moment be stated with any certainty. The City Commissioners are taking their present action with a view to having necessary repairs made to the statue and the pedestal rebuilt in a substantial way. The police authorities would like to have the statue removed altogether, for the reason that it obstructs traffic in the Dame Street area. It is certain that for some time to come, after today, the statue will be absent from its accustomed position . . . The horse is already in slings and today will be lowered and taken away.'

7   Although the explosion had been quite ineffective the Corporation decided to remove the statue as it was considered a traffic hazard. The IRA completed its unfinished business by entering the builders' yard where the statue lay and cutting off the Prince of Orange's head, which they took away. They thus fulfilled the intention plotted by the Defenders 133 years previously. Some months later, the one other monument to William III in the Irish Free State, located at Boyle, County Roscommon, was also destroyed in an explosion. In an interesting postscript the statue provoked further discussion long after it had been removed from the streets of the capital when the question of its re-erection arose during discussions about the commemoration of the 1798 rebellion in 1997. The Orange Order wrote to Dublin Corporation in March asking it to replace the statue, see *The Irish Times*, 6 Mar. 1997.

8   'Vagaries of Dublin', *The Sunday Times*, 6 June 1937.

9   Bodkin was Director of the National Gallery, 1927–1934. Kelly points out that he left Ireland in some bitterness because all his efforts to improve conditions at the gallery had been in vain. She suggests that his attitudes and cultural tastes were at odds with the ardent nationalism of his time, hence his departure must have been a cause of much relief and some rejoicing among the civil servants. See A. Kelly, 'Van Nost's equestrian statue of George I', *Irish Arts Review* 11 (1995): 103–7.

10   Barber Institute, Thomas Bodkin to Robert Atkinson, 27 May 1937, quoted in Kelly, 'Van Nost', p. 106.

11   *Ibid.* p. 106.

12   *Ibid.*

13   *Ibid.*

14   *Ibid.*

15   Minutes of the general purposes committee, Dublin Corporation, 28 Sept. 1937.

16   Public art in Birmingham, Information Sheet, no. 17.

17   Two requests for the statue were made by the Royal Dublin Society, which had received its charter from George II in the 1730s, but both were refused. Some indication that the statue might come under attack was evident in the disturbances in the city a few nights earlier when 'the windows of two Dublin shops were smashed. One was the window of the Dublin Antique Galleries . . . where a small statue of Queen Victoria . . . [was] on display . . . The other window was . . . in Johnston's Court off Grafton Street where British records and Coronation albums were displayed . . . The Dublin Antique Galleries, formerly the Royal Antique Galleries of Middle Abbey Street in pre-treaty days supplied furnishings for Dublin Castle, the Viceregal Lodge and the RHK for Royal visits to Ireland', *The Irish Times*, 14 May 1937, p. 7.

18   *The Irish Times*, 14 May 1937, p. 7.

19   *The Irish Times*, 27 Dec. 1944, p. 2. See also *Dáil Debates*, 1945, cols. 129–30, where the desire to have the statue removed altogether was expressed by some deputies.

20   *Dáil Debates*, 1932, col. 239.

21   See OPW files: Disposal of Queen Victoria's statue, A 25/16/4/2, A 25/16/5, A 25/16/2/1.

22   *The Star*, 10 Aug. 1929, p. 1.

23   *The Star*, 17 Aug. 1929, p. 4.

24   *The Irish Times*, 19 Aug. 1929, p. 8.

25   *Dáil Debates*, 1930, col. 1223, see also further calls for its removal, 1932, col. 239.

26   *Irish Press*, 8 Feb. 1933.

27   *Dáil Debates*, 3 Feb. 1937, col. 7.

28   *IrB*, 25 Dec. 1937, p. 1118.

29   *Dublin Evening Herald*, 24 Oct. 1938, p. 5.

30   *Minutes*, 1943, nos 195, 246.

31   *Dáil Debates*, 1948, col. 1786. See also, 1949, col. 1311, 1947, col. 537.

32   *The Irish Times*, 23 July 1948, p. 3. It was later moved to museum storage in Daingean, Co. Offaly.

33   These requests are recorded in the files of the OPW.

34   See *Dáil Debates*, 1949, cols 544–6 and T. Bodkin *Report on the arts in Ireland* (Dublin, 1949). He notes in ch. 10, 'The preservation of national monuments and sites', that 'No one could dispute the propriety of removing Queen Victoria's monument from the front of Leinster House, nor regret the transfer of the ugly statue of Her Britannic majesty to London, Ontario, or Peterborough in Canada; but many will wish that at least the beautiful figure of Fame which adorned the back of that monument should be put on view somewhere in the native country of its sculptor, John Hughes', p. 52.

35   OPW file: Queen Victoria statue, A/25/16/4, vol. 1, no. 137. A US businessman later showed an interest in purchasing the monument, while in 1966 as part of the fiftieth anniversary commemoration of the 1916 Rising it was suggested that 'it would be an opportune time for the department to dispose of all these old relics of the past.' Further requests were received throughout the 1980s from Thunder Bay, Ontario, and Halifax, Nova Scotia.

36   QVB promotional literature, n.d.g.

37   OPW, Queen Victoria Statue, no. 270.

38   *Ibid.*, no. 273.

39   *Ibid.*, no. 275.

40   *Ibid.*, no. 284. The QVB authorities did not receive the entire monument, however. They only requested and received the figure of the Queen. The other parts, namely the statue of Ceres representing Hibernia at peace, the statue of a dying soldier representing Ireland at war, an angel figure referred to as 'fame', remained in storage at Daingean, County Offaly.

41   Not all, however, are content with the erection of the monument in Sydney's city centre. The *Sun-Herald* commented on St Patrick's Day 1996 that, 'This ugly relic of British colonial rule in Ireland was dispatched down-under because nobody in the Emerald Isle wanted it. Its presence in Sydney is an offence to Irish Australians and an insult to Australian Australians', *Sun-Herald*, 17 Mar. 1996, p. 22.

42   *IrB*, 11 Mar. 1922, p. 1.

43   Cited in P. Henchy, 'Nelson's Pillar', in *Dublin Historical Record* 10 (1948–9): 53–63, p. 62.

44   *IrB*, 25 July 1925, p. 613.

45   *IrB*, 3 Apr. 1926, p. 1.

46   *IrB*, 21 Aug. 1926, p. 604.

47   *Irish Press*, 9 Dec. 1931.

48   Minutes of the Corporation of Dublin, 1948, no. 219.

49   *Ibid.*

50   Minutes, 1954, no. 186.

51   *Ibid.*

52   Minutes, 1955, no. 82.

53   Minutes, 1955, no. 82.

54   Minutes, 1955, no. 82.

55   Minutes, 1956, no. 73.

56   Cited in M. O' Riain, 'Nelson's Pillar: a controversy that ran and ran', *History Ireland* 6, 4 (1998): 21–5, p. 24.

57   *The Irish Times*, 8 Mar. 1966, p. 1.

58   *The Irish Times*, 9 Mar. 1966, p. 4.

59   *Seanad Debates*, 1969, cols 915–16.

60   *Evening Press*, 8 Mar. 1966.

61   *The Irish Times*, 27 Aug. 1956, p. 7.

62   On Christmas Eve 1944, the head of the figure of Lord Gough was removed, while paint was poured over it in Oct. 1956. See OPW: Gough Junction, erection of a monument, P 2/74/3 and Gough Monument, Malicious damage, P 2/74/ 2.

63   *The Irish Times*, 6 Nov. 1956.

64   *The Irish Times*, 24 July 1957.

65   *The Irish Times*, 29 July 1958, p. 1.

66   *Ibid.*

67   *Ibid.*

68   *The Irish Times*, 27 Aug. 1958, p. 7.

69   *Ibid.*

70   While his proposal was given consideration by the OPW, he subsequently declined the offer, see OPW, P 2/74/2.

71   See *The Irish Times*, 20, 22 Aug., 23 Sept. 1986.

72   See D. Fitzpatrick, 'Commemoration in the Irish Free State: a chronicle of embarrassment', in I. McBride (ed.), *History and memory in Modern Ireland* (Cambridge, 2001), pp. 184–204.

73   'Memorials to Irish patriots: erection on state property', 7 Mar. 1939 (NA, CAB2/2 2nd gov. cabinet).

Chapter 8 Street naming and nation building in Dublin

1   Minutes, 1920, no. 758.

2   RPDCD, 1921, vol. I, no. 71.

3   *Ibid.*

4   *Ibid.*

5   *Ibid.*

6   *Ibid.*

7   *Ibid.*

8   *Ibid.*

9   Minutes, 1922, no. 35.

10   RPDCD, 1921, vol. II, no. 209.

11   Minutes, 1921, no. 35.

12   RPDCD, 1922, vol. II, no. 327.

13   Minutes, 1923, no. III.

14   Minutes, 1924, no. 278; see also Minutes, 1933, no. 54.

15   Minutes, 1924, no. 296.

16   Minutes, 1924, no. 297. In the same year M'Clean's Lane changed to Meade's Terrace as there existed another street with the same name, Minutes, 1924, no. 368.

17   RPDCD, 1932, no. 4.

18   RPDCD, 1932, no. 40.

19   Other names included Stannaway, Devenish, Leighlin, Kildare, Durrow, Kilfenora, Monasterboice, Lismore Tonguefield, Slane, Downpatrick, and Saul Road, see Minutes, 1934, no. 393.

20   Minutes, 1936, no. 96.

21   *Dáil Debates*, 1944, cols. 1517–18.

22   *Dáil Debates*, 1945, col. 672. The Bill dealt with 34 different statutes.

23   *Dáil Debates*, 1945, col. 581.

24   *Dáil Debates*, 1945, col. 582.

25   *Dáil Debates*, 1945, col. 674.

26   *Ibid.*

27   *Dáil Debates*, 1945, col. 675.

28   *Dáil Debates*, 1946, col. 24. For earlier legislation see the Public Health Acts amendment Act 1907 (7 Dew vii, c. 53), s. 21. This required the consent of two thirds in number and value of the ratepayers in any street to alter its name.

29   *Dáil Debates*, 1945, col. 582.

30   T. Brown, 'Architecture in Independent Ireland', in Government of Ireland, *Building for Government* (Dublin, 1999), pp. 21–5.

31   Names such as downs, dean, copse and spinney began to appear. These names are 'not merely pretentious, but often force on to have recourse to the dictionary, as they are not terms in normal use or readily understood in Ireland . . . and heralded the transformation of Ireland's urban namescape', see L. Mac Mathúna, 'Urban placenames; streets and districts', in A. Ó Maolfabhail (ed.), *The placenames of Ireland in the third millennium* (Dublin, 1992), pp. 53–70, p. 68.

Chapter 9   The power of space

1   D. Ley and J. Duncan, 'Epilogue', in J. Duncan and D. Ley (eds), *Place/culture/representation* (London, 1993), pp. 329–34, p. 329.

2   E. Soja, *Thirdspace: Journeys to Los Angeles and other real and imagined places* (Oxford, 1996), p. 46. See also E. Soja, *Postmodern geographies: The reassertion of space in critical social theory* (London, 1989).

3   R.A. Dodgshon, *Society in time and space* (Cambridge, 1998), p. 4.

4   C. Harris, 'Power, modernity, and historical geography', *AAAG* 81 (1991): 671–83, p. 678. See also M. Foucault, *Power/Knowledge* (Brighton, 1980).

5   H. Lefebvre, *The production of space* (Oxford, 1991).

6   See M. Gottdiener, 'Culture, ideology, and the sign of the city', in M. Gottdiener and A. P. Lagopoulos, *The city and the sign* (New York, 1986), p. 203; D. Harvey, *The limits to capital* (Oxford, 1982); D. Harvey, *Consciousness and the urban experience* (Oxford, 1985); D. Harvey, *The condition of postmodernity: an inquiry into the origins of cultural change* (Oxford, 1989). See also A. Madanipour, *Design of urban space: an inquiry into a socio-spatial process* (Chichester, 1996). On space as a crucial element within contemporary cultural, literary and historical studies in Ireland see G. Smyth, *Space and the Irish cultural imagination* (Basingstoke, 2001).

7   J. Duncan and D. Ley, 'Epilogue', p. 329. See also D. Gregory and D. Ley, 'Culture's geography – editorial', *E&P,D* 6 (1988): 115–16.

8   Dublin Corporation, *A new monument for O'Connell Street. International Competition* (Dublin, 1998), p. 3.

9   I. McBride, 'Memory and national identity in modern Ireland', in I. McBride, (ed.), *History and memory in modern Ireland* (Cambridge, 2001), pp. 1–42, p. 1.

10   While contemporary public monuments can be described as largely apolitical, tending towards public art, a number of more political monuments have been erected. The Famine monument erected in 1996 on the banks of the River Liffey and adjacent to the new International Financial Services Centre is a case in point, as is the James Connolly monument close by.

11   As was reported in *The Irish Times*, 'The ceremony outside the GPO, which involved the reading of the Proclamation and the hoisting of the national flag, lasted 15 minutes and was attended by some members of the Government and a small number of TDs and senators. Five veterans of the Rising with members of their families also attended', 1 Apr. 1991, p. 1. Other commemorations were held at Kilmainham Jail, at Enniscorthy, County Wexford, Pearse's Cottage at Rosmuc, County Galway, in Belfast, Limerick and Kerry and at the republican plot at Glasnevin.

12   *The Irish Times*, 20 Dec. 1978.

13   *The Irish Times*, 22 Aug. 1979.

14   *Irish Independent*, 22 Aug. 1979.

15   *The Irish Times*, 20 Nov. 1979.

16   *Ibid.*

17   *The Irish Times*, 11 Feb. 1988.

18   H. Murray, The Pillar project, *Irish Architect*, 1988, Nov./Dec., pp. 20–1; J. O'Regan (ed.), *A monument in the city* (Cork, 1998).

19   *The Irish Times*, 30 July 1998.

20   *The Irish Times*, 26 Nov. 1998.

21   *Ibid.*

22   *Ibid.*

23   *Ibid.*

24   Dublin Corporation, *A new monument*, p. 3.

25   *The Irish Times*, 22 Apr. 2001.

26   *The Irish Times*, 13 Oct. 2001.

27   *The Irish Times*, 8 Oct. 2001.

28   See reports in *The Irish Times*, 10, 11 12 Nov. 1998.

Bibliography

For abbreviations used in the Bibliography, see p. 243.

AAI, *Public works. The Architecture of the Office of Public Works 1831–1987* (Dublin, 1987).

F. Aalen and K. Whelan (eds), *Dublin city and county, from prehistory to present* (Dublin, 1992).

T. Aase, 'Symbolic space: representations of space in geography and anthropology', *Geografiska Annaler* 76, B, 1 (1994): 51–8.

P. Abercrombie, 'The Dublin town plan', *Studies* 31 (1942): 155–60.

P. Abercrombie, S. Kelly and A. Kelly, *Dublin of the future* (Dublin, 1922).

P. Abercrombie, S. Kelly and M. Robertson, *Dublin sketch development plan* (Dublin, 1941).

D. Ades, T. Benton, D. Elliot and I. Boyd-White (eds), *Art and power: Europe under the Dictators, 1930–45* (London, 1995).

M. Aghulon, *Marianne into battle: republican imagery and symbolism in France, 1789–1880* (Cambridge, 1981).

J. A. Agnew and J. S. Duncan (eds), *The power of place: bringing together geographical and sociological imaginations* (London, 1989).

J. A. Agnew, D. N. Livingstone and A. Rogers (eds), *Human geography: an essential anthology* (Oxford, 1996).

J. A. Agnew, J. Mercer and D. Sopher (eds), *The city in a cultural context* (London, 1984).

M. Allen, 'Jerome Connor – One', *The Capuchin Annual* (1963): 347–68; 'Jerome Connor – Two', *The Capuchin Annual* (1964): 353–69; 'Jerome Connor – Three', *The Capuchin Annual* (1965): 365–87.

N. Al Sayyad (ed.), *Forms of dominance: on the architecture and urbanism of the colonial enterprise* (London, 1992).

P. Alter, 'Symbols of Irish nationalism', *Studia Hibernica* 14 (1974): 14–123.

A. Åman, *Architecture and ideology in eastern Europe during the Stalin era* (London, 1992).

B. Anderson, *Imagined communities: reflections on the origins and spread of nationalism* (London, 1983).

——'Cartoons and monuments: the evolution of political communication under the New Order', in D. Jackson and L. Pye (eds), *Political power and communications in Indonesia* (Berkeley, 1987).

K. Anderson, 'Chinatown as a public nuisance: the power of place in the making of a racial category', *AAAG* 77 (1987): 580–98.

——'Cultural hegemony and the race definition process in Chinatown, Vancouver, 1880–1980', *E&P, D* 6 (1988): 127–49.

K. Anderson and F. Gale (eds), *Inventing places: studies in cultural geography* (Melbourne, 1992).

——*Cultural geographies* (London, 1999).

R. Argenbright, 'Remaking Moscow: new places, new selves', *Geographical Review*, 89, 1 (1999): 1–22.

G. J. Ashworth, *On tragedy and Renaissance* (Groningen, 1993).

——'The conserved European city as cultural symbol: the meaning of the text', in B. Graham (ed.), *Modern Europe: place, culture and identity* (London, 1998), pp. 261–86.

G. J. Ashworth and P. Howard, *European heritage planning and management* (Exeter, 1999).

D. Atkinson and D. E. Cosgrove, 'Urban rhetoric and embodied identities: city, nation and empire at the Vittorio Emanuele II monument in Rome, 1870–1945', *AAAG* 88, 1 (1998): 28–49.

M. Azaryahu, 'Street names and political identity: the case of East Berlin', *Journal of Contemporary History* 21 (1988): 581–604.

——'Renaming the past: changes in "city text" in Germany and Austria, 1945–1947', *History and Memory* 22 (1990): 32–53.

——'The power of commemorative street names', *E&P,D* 14, 3 (1996): 311–30.

——'The spontaneous formation of memorial space: the case of Kikar Rabin, Tel Aviv', *Area* 28, 4 (1996): 501–13.

——'German reunification and the power of street names, *PG* 16, 6 (1997): 479–93.

——'McDonald's or Golani Junction? A case of a contested place in Israel', *Professional Geographer* 51, 4 (1999): 481–92.

B. A. Badcock, 'Looking-glass views of the city', *PHG* 20, 1 (1996): 91–7.

A. R. H. Baker, 'Introduction: on ideology and landscape', in Baker and G. Biger (eds), pp. 1–14.

A. R. H. Baker and G. Biger (eds), *Ideology and landscape in historical perspective* (Cambridge, 1992).

M. Bannon (ed.), *The emergence of Irish planning 1880–1920* (Dublin, 1985).

——'The capital of the new state', in Cosgrove (ed.), pp. 133–50.

——(ed.), *Planning: the Irish experience, 1920–1988* (Dublin, 1989).

C. Bardon and J. Bardon, *If ever you go down to Dublin town: a historic guide to the city's street names* (Belfast, 1988).

T. J. Barnes and J. Duncan (eds), *Writing worlds: discourse, text and metaphor in the representation of the landscape* (London, 1992).

T. J. Barnes and D. Gregory (eds), *Reading human geography: the poetics and politics of inquiry* (London, 1997).

C. Barrett, 'Irish nationalism and art 1800–1921', *Studies* (1975): 393–409.

T. Bartlett et al (eds), *Irish studies: a general introduction* (Dublin, 1988), 'What ish my nation? Themes in Irish history, 1550–1850', pp. 44–59.

A. Becker, J. Olley and W. Wang (eds), *Twentieth century architecture: Ireland* (Munich, 1997).

J. Bell, 'Redefining national identity in Uzbekistan', *Ecumene* 6, 2 (1999): 183–207.

B. Bender (ed.), *Landscape: politics and perspectives* (Oxford, 1993); 'Introduction: landscape – meaning and action', pp. 1–18.

D. Bennet, *A Dublin anthology* (Dublin, 1994).

L. D. Berg and R. A. Kearns, 'Naming as norming: 'race', gender, and the identity politics of naming places in Aoteraoa/New Zealand', *E&P,D* 14 (1996): 99–122.

H. Bhabha (ed.), *Nation and naviation* (London, 1990).

S. Bhreathnach-Lynch, 'Albert Power, RHA', *Irish Arts Review* 7 (1990–1): 111–14.

——'Face value: commemoration and its discontents', *Circa* 65 (1993): 32–7.

——'"Executed": the political commissions of Albert G. Power', *Éire–Ireland* Spring (1994): 44–60.

——'John Hughes: The Italian Connection', *Irish Arts Review* 10 (1994): 195–201.

——'The art of Albert Power, 1881–1945: a sculptural legacy of Irish Ireland', in B. P. Kennedy and R. Gillespie (eds), *Ireland: art into history* (Dublin, 1996), pp. 118–32.

I. S. Black, 'Symbolic capital: the London and Westminster Bank headquarters, 1836–38', *Landscape Research* 21, 1 (1996): 55–72.

T. Bodkin, *Report on the arts in Ireland* (Dublin, 1949).

J. Bodnar, 'Public memory in an American city: commemoration in Cleveland', in Gillis (ed.), pp. 74–89.

A. Boime, *Hollow icons: the politics of sculpture in nineteenth century France* (Kent, Ohio and London, 1987).

W. Bolger and B. Share, *And Nelson on his pillar, 1808/1966: a retrospective record* (Dublin 1976).

L. Bondi, 'Gender symbols and urban landscapes', *PHG* 16, 2 (1992): 157–70.

G. Boyce, *Nationalism in Ireland* (London, 1991).

——'Ireland and the First World War', *History Ireland* 2, 3 (1994): 48–53.

Lt. Col. Boydell, *The Irish National War Memorial: its meaning and purpose. A permanent reminder of the splendor of Irish valour and a visible expression of our unparalleled national unity*, NLI unpublished account.

M. C. Boyer, *The city of collective memory* (Massachusetts, 1994).

J. Brady and A. Simms (eds), *Dublin through space and time* (Dublin, 2001).

T. Brown, *Ireland: a social and cultural history 1922–1985* (London, 1985).

——'Architecture in Independent Ireland', in Government of Ireland, *Building for Government* (Dublin, 1999), pp. 21–5.

J. Burgess, 'The future for landscape research', *Landscape Research* 21, 1 (1996): 5–12.

R. M. Butler, 'The reconstruction of O'Connell Street, Dublin', *Studies* Dec. (1916): 570–6.

A. Buttimer, *Values in geography*, Washington DC, AAG, Commission on College Geography, no. 24, (Washington DC, 1974).

——'Reason, rationality and human creativity', *Geografiska Annaler* 61 B 1979: 43–9.

——'Geography, humanism and global concern', *AAAG* 80 (1990): 10–33.

H. Campbell, 'Religion and the city: the Catholic Church in Dublin 1691–1878', *Urban Design Studies* 3 (1993): 1–24.

——'Irish identity and the architecture of the new state', in Becker et al. (eds), pp. 82–5.

S. Cassidy, 'The cultural identity of Ireland in literature and the built environment', in Becker et al. (eds), pp. 72–7.

R. Cathcart, *William III and Ireland* (Dublin, 1990).

M. Cavalcanti, 'Urban reconstruction and autocratic regimes: Ceausescu's Bucharest in its historic context', *Planning Perspectives* 12 (1997): 71–109.

A. J. Christopher, '"The second city of the Empire": colonial Dublin, 1911', *JHG* 23, 2 (1997): 151–63.

Civics Institute of Ireland, *The Dublin civic survey* (Liverpool, 1925).

Civic Week Council, *Official handbook, 1927* (Dublin, 1927).

——*Official handbook, 1929* (Dublin, 1929).

P. Cloke, C. Philo and D. Sadler (eds), *Approaching human geography: an introduction to contemporary theoretical debates* (London, 1991).

S. B. Cohen and N. Kliot, 'Placenames in Israel's ideological struggle over the administered territories', *AAAG* 82, 4 (1992): 653–80.

W. Cohen, 'Symbols of power: statues in nineteenth century provincial France', *Comparative Studies in History and Society* 31 (1989): 491–513.

H. Conway and R. Roenisch, *Understanding architecture* (London, 1994).

S. J. Connolly, 'Culture, identity and tradition', in B. Graham (ed.), pp. 43–63.

P. Coones, 'Landscape geography', in A. Rogers, H. Viles and A. Goudie (eds), *The student's companion to geography* (Oxford, 1992), pp. 70–6.

A. Cosgrove (ed.), *Dublin through the ages* (Dublin, 1988).

D. E. Cosgrove, 'Place, landscape and the dialectics of cultural geography', *Canadian Geographer* 22 (1978): 66–72.

——'John Ruskin and the geographical imagination', *Geographical Review* 69 (1979): 43–62.

——'Towards a radical cultural geography', *Antipode* 15 (1983): 1–11.

——*Social formation and symbolic landscape* (London, 1984).

——'Prospect, perspective and the evolution of the landscape idea', *TIBG* 10 (1985): 45–62.

——'The terrain of metaphor: cultural geography, 1988–89', *PIHG* 13 (1989): 566–75.

——'Geography is everywhere: culture and symbolism in human landscapes', in Gregory and Walford (eds), pp. 118–35.

——'Environmental thought and action: pre-modern and post-modern', *TIBG* 15 (1990): 344–58.

——'Orders and a new world: cultural geography, 1990–1991', *PHG* 16 (1992): 72–80.

——'On "The reinvention of cultural geography" by Price and Lewis. Commentary', *AAAG* 83, 3 (1993): 515–16.

——'Worlds of meaning: cultural geography and the imagination', in Foote, Hughill, Mathewson and Smith (eds), pp. 387–98.

——'Ideas and culture: a response to Don Mitchell', *TIBG* 23, 3 (1996): 574–5.

——*The Palladian landscape* (Leicester, 1996).

——'Windows on the city', *Urban Studies* 33, 8 (1996): 1495–8.

——'Cultural landscapes', in T. Unwin (ed.) (1998), pp. 65–81.

D. E. Cosgrove and S. Daniels (eds), *The iconography of landscape* (Cambridge, 1988).

D. E. Cosgrove and M. Domosh, 'Author and authority: writing the new cultural geography', in Duncan and Ley (eds), pp. 25–38.

D. E. Cosgrove and P. Jackson, 'New directions in cultural geography', *Area* 19, 2 (1987): 95–101.

H. Cotter, 'The reconstruction of the Four Courts following its destruction in 1922', unpublished BArch dissertation, School of Architecture, UCD, 1989.

M. Craig, *Dublin 1660–1860* (Dublin, 1982).

M. Crang, *Cultural geography* (London, 1998)

C. P. Curran, 'Comments on the foregoing article, No. 4', *Studies* (1942): 167–70.

Curriculum Development Unit, *Divided city: portrait of Dublin 1913* (Dublin, 1978).

M. E. Daly, *Dublin: The deposed capital. A social and economic history: 1850–1900* (Cork, 1985).

——'A tale of two cities: 1860–1920', in A. Cosgrove (ed.), pp. 113–32.

S. Daniels, 'Arguments for a humanistic geography', in R. J. Johnston (ed.), *The future of geography* (London, 1985), pp. 143–58.

——'The political iconography of woodland in later Georgian England', in Cosgrove and Daniels (eds), pp. 43–82.

——'Marxism, culture and the duplicity of landscape', in R. Peet and N. Thrift (eds), *New models in geography* (London, 1989), pp. 196–220.

——*Fields of vision: landscape imagery and national identity in England and the United States* (London, 1992).

S. Daniels and D. E. Cosgrove, 'Introduction' in Cosgrove and Daniels (eds), pp. 1–10.

——'Spectacle and text: landscape metaphors in cultural geography', in Duncan and Ley (eds), pp. 57–77.

A. de Blacam, 'James Clarence Mangan', *The Capuchin Annual* (1949): 153–83.

J. W. de Courcy, *The Liffey in Dublin* (Dublin, 1996).

A. Denson, *John Hughes, sculptor, 1865–1941* (Kendal, 1969).

——'John Hughes, RHA, 1865–1941', *The Capuchin Annual* (1975): 126–37.

Department of Arts, Culture and the Gaeltacht, *Developing a government policy on architecture: a proposed framework and discussion of issues* (Dublin, 1996).

Department of External Affairs, *Cuimhneachán 1916, Commemoration: a record of Ireland's commemoration of the 1916 Rising* (Dublin, 1966).

F. E. Dixon, 'The portrait statues of Dublin', *Dublin Historical Record* 7, 4 (1945): 155–60.

R. A. Dodghson, *Society in time and space* (Cambridge, 1998).

M. Domosh, 'The symbolism of the skyscraper: case studies of New York's first tall buildings', *Journal of Urban History* 14 (1988): 321–45.

——'A method for interpreting landscape: a case study of the New York World building', *Area* 21, 4 (1989): 347–55.

——'New York's first skyscrapers: conflict in the design of the American commercial landscape', *Landscape* 30 (1989): 34–8.

——'Controlling urban form: the development of Boston's Back Bay', *JHG* 18, 3 (1992): 288–306.

——'Urban imagery', *Urban Geography* 13, 5 (1992): 475–80.

——'The symbolism of the skyscraper: case studies of New York's first tall buildings', in Foote, Hughill, Mathewson, and Smith (eds), pp. 48–63.

——'Corporate cultures and the modern landscape of New York City', in F. Gale and K. Anderson (eds), *Inventing places: studies in cultural geography* (Melbourne, 1995), pp. 73–86.

——'A "feminine building"? Relations between gender ideology and aesthetic ideology in turn of the century America,' *Ecumene*, 1996, 3, 3, pp. 305–324.

K. Dovey, 'Corporate towers and symbolic capital', *Environment and Planning B: Planning and Design* 19 (1992): 173–88.

F. Driver and D. Gilbert (eds), *Imperial cities: landscape, display and identity* (Manchester, 1999); 'Imperial cities, overlapping territories, intertwined histories', pp. 1–17.

Dublin Corporation, *Official guide to the city of Dublin* (Dublin, 1953).

——*O'Connell Street integrated area action plan* (Dublin, 1998).

—— *A new monument for O'Connell Street: international competition* (Dublin, 1998).

G. Duffy, 'The Dublin town plan – comment', *Studies* 31 (1942): 161–3.

J. Duncan, 'The superorganic in American cultural geography', *AAAG* 70 (1980): 181–98.

——'Changes in authority and meaning under three cultural paradigms in Kandy, Sri Lanka', in D. Saile (ed.), *Architecture in cultural change* (Lawrence, 1985).

——'Review of urban imagery: urban semiotics', *Urban Geography* 8 (1987): 473–83.

—— *The city as text: the politics of language interpretation in the Kandyan kingdom* (Cambridge, 1988).

——'The power of place in Kandy, Sri Lanka: 1780–1980', in Agnew and Duncan (eds), pp. 185–201

——'Sites of representation: place, time and the discourse of the other', in Duncan and D. Ley (eds), pp. 39–56.

——'Representing power: the politics and poetics of urban form in the Kandyan Kingdom', in Duncan and D. Ley (eds), pp. 232–48.

——'The politics of landscape and nature, 1992–93', *PHG* 18, 3 (1994): 361–70.

——'After the civil war: reconstructing cultural geography as heterotopia', in Foote, Hughill, Mathewson, and Smith (eds), pp. 401–8.

——'On "The reinvention of cultural geography" by Price and Lewis. Commentary', *AAAG* 83, 3 (1993): 517–19.

——'Landscapes of the self/landscapes of the other(s): cultural geography 1991–92', *PHG* 17, 3 (1993): 367–77.

——'Landscape geography 1993–1994', *PHG* 19, 3 (1995): 414–22.

J. Duncan and N. Duncan, '(Re)reading the landscape', *E&P,D* 6 (1988): 117–26.

——'Reconceptualising the idea of culture in geography: a reply to Don Mitchell', *TIBG* 21, 3 (1996): 576–9.

J. Duncan and D. Ley (eds), *Place/culture/representation* (London and New York, 1993); 'Introduction: representing the place of culture', pp. 1–24.

K. M. Dunn, 'Cultural geography and cultural policy', *Australian Geographical Studies* 35, 1 (1997): 1–11.

C. Earle, K. Matthewson and M. Kenzer (eds), *Concepts in human geography* (London, 1996).

T. Edensor, 'National identity and the politics of memory: remembering Bruce and Wallace in symbolic space', *E&P,D* 29 (1997): 175–94.

J. N. Entrikin, 'Contemporary humanism in geography', *AAAG* 66 (1976): 615–32.

—— *The betweeness of place: towards a geography of modernity* (London, 1991).

J. Eyles, 'The geography of everyday life', in R. Walford and D. Gregory (eds), *Horizons in human geography* (London, 1989), pp. 102–17.

J. Eyles and W. Peace, 'Signs and symbols in Hamilton: an iconlogy of Steeltown', *Geografiska Annaler* 72, B (1990): 73–88.

P. Fagan, 'The population of Dublin in the eighteenth century with particular reference to the proportions of Protestants and Catholics', *Eighteenth Century Ireland* 6 (1991): 121–56.

B. Fallon, *Irish Art, 1830–1990* (Belfast, 1994).

——*An age of innocence: Irish culture 1930–1960* (Dublin, 1998).

P. Ferguson, 'Reading city streets', *French Review* 61 (1988): 386–97.

D. Fitzpatrick (ed.), *Ireland and the First World War* (Dublin, 1986).

——*The two Irelands, 1912–1939* (Oxford, 1998).

——'Commemoration in the Irish Free State: a chronicle of embarrassment', in I. McBride (ed.), pp. 184–204.

K. E. Foote, P. J. Hughill, K. Mathewson, and J. Smith (eds), *Re-reading cultural geography* (Austin, 1994).

K. E. Foote, A. Tóth and A. Árvay, 'Hungary after 1989: inscribing a new past on place', *Geographical Review* 90, 3 (2000): 301–34.

R. Foster, *Modern Ireland 1600–1972* (London, 1988).

——'Remembering 1798', in R. F. Foster, *The Irish story* (London, 2001), pp. 211–34.

M. Fraser, 'Public building and colonial policy in Dublin, 1760–1800', *Architectural History* 28 (1985): 102–22.

N. Fyfe (ed.), *Images of the street: planning, identity, and control in public space* (London, 1998).

P. F. Garnett, 'The Wellington testimonial', *Dublin Historical Record*, 8 (1955–6): 48–61.

T. Garvin, *1922: the birth of Irish democracy* (Dublin, 1996).

C. Geertz, *The interpretation of cultures* (New York, 1973).

L. Gibbons, *Transformations in Irish culture* (Cork, 1996).

J. T. Gilbert, *History of the city of Dublin*, 3 vols (Dublin, 1854–9, reprinted Shannon, 1972).

——*Calendar of the Ancient Records of Dublin* (Dublin, 1889–1922).

J. R. Gillis (ed), *Commemorations: The politics of national identity* (Princeton NJ, 1994); 'Memory and identity: the history of a relationship', pp. 3–24.

L. F. Giron, 'Comments on the foregoing article, no. 2', *Studies*, 1942, pp. 163–4.

Glasnevin Cemeteries Group, *Glasnevin cemetery: an historic walk* (Dublin, 1997).

I. Golomstock, *Totalitarian art in the Soviet Union, the Third Reich, Fascist Italy and the People's Republic of China* (New York, 1990).

M. Gorham, *Dublin from old photographs* (London, 1972).

J. Goss, 'The built environment and social theory: towards an architectural geography', *Professional Geographer* 40, 4 (1988): 392–402.

M. Gottdiener, 'Urban semiotics', in J. Pipkin, M. LaGory and J. Blau (eds), *Remaking the city* (Albany, 1983).

——*Postmodern semiotics* (Oxford, 1995).

M. Gottdiener and A. Lagopoulos (eds), *The city and the sign* (New York, 1986).

Government of Ireland, *Building for Government* (Dublin, 1999).

B. Graham (ed.), *In search of Ireland* (London, 1997).

B. Graham, 'No place of the mind: contested Protestant representations of Ulster', *Ecumene* 1, 3 (1994): 257–82.

——*Modern Europe: place, culture and identity* (London, 1998).

——The past in place: historical geographies of identity', in Graham and Nash (eds), pp. 70–99.

B. Graham, G. J. Ashworth and J. E. Tunbridge, *A geography of heritage* (London, 2000)

B. Graham and C. Nash (eds), *Modern historical geographies* (Harlow, 2000)

E. Graham, 'Postmodernism and the possibility of a new human geography', *Scottish Geographical Magazine* 111, 3 (1995): 175–8.

Greater Dublin Reconstruction Movement, *Greater Dublin reconstruction scheme described and illustrated* (Dublin, 1922).

D. Gregory, *Ideology, science and human geography* (London, 1978).

D. Gregory and D. Ley, 'Culture's geographies', *E&P, D* 6 (1988): 155–6.

D. Gregory, R. Martin and G. Smith (eds), *Human geography* (London, 1994).

D. Gregory and R. Walford (eds), *Horizons in human geography* (London, 1989).

C. L. Griswold, 'The Vietnam veteran's memorial and the Washington Mall: philosophical thoughts on political iconography', *Critical Inquiry* 12 (1986): 689.

P. Gruffudd, 'Remaking Wales: nation-building and the geographical imagination, 1925–50', *PG* 14, 3 (1995): 219–39.

H. E. Gulley, 'Women and the lost cause: preserving Confederate identity in the American Deep South', *JHG* 19 (1993): 125–41.

T. Haarni, 'Modern art and revenge of the military hero: new planning practices and the geography of resistance in central Helsinki', *Nordisk Samhällsgeografisk Tidskrift* 20 (1995): 116–30.

P. Hall, *Cities of tomorrow* (Oxford, 1988).

——*Cities in civilisation* (London, 1998)

T. Hall, *Planning Europe's cities, aspects of nineteenth century urban development* (London, 1997).

R. Handler, 'Is "Identity" a useful cross-cultural concept?' in Gillis (ed), pp. 27–40.

F. Hansen 'Approaching human geography: towards new approaches in human geography', *Geografiska Annaler* 76 B, 3 (1994): 211–16.

C. Harris, 'Power, modernity and historical geography', *AAAG* 81, 4 (1991): 671–83.

D. Harvey, 'Monument and myth', *AAAG* 69 (1979): 362–81.

——*The limits to capital* (Oxford, 1982).

——(ed.), *Consciousness and the urban experience* (Oxford, 1985).

——*The urban experience* (Oxford, 1985).

——*The condition of postmodernity: an inquiry into the origins of cultural change* (Oxford, 1989).

——*Spaces of hope* (Edinburgh, 2000)

V. Hastaglou-Martinidis, K. Kaflouka and N. Papanichos, 'Urban modernisation and national renaissance; town planning in 19th century Greece', *Planning Perspectives* 8 (1993): 427–69.

M. Heffernan, 'For ever England: the Western Front and the politics of remembrance in Britain, *Ecumene* 2, 3 (1995): 293–323.

——'War and the shaping of Europe', in B.Graham (ed.), (1998), pp. 89–115.

——*The meaning of Europe: geography and geopolitics* (Oxford, 1998).

P. Henchy, 'Nelson's Pillar', *Dublin Historical Record* 10 (1948–9): 53–63.

J. Hill, *Irish public sculpture* (Dublin, 1998).

E. Hobsbawm, 'Foreword', in D. Ades et al. (eds).

——'Mass producing traditions: Europe, 1870–1914', in Hobsbawm and Ranger (eds), pp. 263–309.

E. Hobsbawm and T. Ranger (eds), *The invention of tradition* (Cambridge, 1983).

D. Holdsworth, 'Architectural expressions of the Canadian national state', *Canadian Geographer* 30, 2 (1986): 167–71.

S. Howe, *Ireland and empire* (Oxford, 2000).

P. J. Hughill and K. E. Foote, 'Re-reading cultural geography', in Foote et al. (eds), pp. 9–23.

W. Hung, 'Tiananmen Square: a political history of monuments', *Representations* 35 (1991): 84–104.

K. S. Inglis, 'Entombing unknown soldiers: from London and Paris to Baghdad', *History and Memory* 5, 2 (1993): 7–49.

P. Jackson, 'A plea for cultural geography', *Area* 12 (1980): 110–13.

——*Maps of Meaning: an introduction to cultural geography* (London, 1989).

——'Mapping meanings: a cultural critique of locality studies', *Environment and Planning A* 23 (1991): 215–28.

——'Berkeley and beyond: broadening the horizons of cultural geography', *AAAG* 83, 3 (1993): 519–20.

——'The idea of culture: a response to Don Mitchell', *TIBG* 21, 3 (1996): 572–3.

——'Geography and the cultural turn', *Scottish Geographical Magazine* 113, 3 (1997): 186–8.

J. Jacobs, 'The city unbound: qualitative approaches to the city', *Urban Studies* 30, 4/5 (1993): 827–48.

——*Edge of empire* (London, 1996).

L. Jakubowska, 'Political drama in Poland: the use of national symbols', *Anthropology Today* 6 (1990): 10–13.

K. Jeffery, *Ireland and the Great War* (Cambridge, 2000)

R. Jeffrey, 'What the statues tell: the politics of choosing symbols in Trivanrum', *Pacific Affairs* 23 (1980): 484–502.

C. Jencks, *Architecture today* (New York, 1982), p. 178.

——*The language of postmodern architecture* (London, 1987).

——(ed.), *The postmodern reader* (London and New York, 1992).

C. Jencks and M. Valentine, 'The architecture of democracy, the hidden tradition', *Architectural Design* 57 (1987): 8–25.

N. C. Johnson, 'Sculpting heroic histories: celebrating the centenary of the 1798 rebellion', *TIBG* 19 (1994): 78–93.

——'Cast in stone: monuments, geography and nationalism', *E&P,D* 13 (1995): 51–65.

——'Nations and peoples', in T. Unwin (ed.), *A European geography* (London, 1998), pp. 85–99.

——'Historical geographies of the present', in Graham and Nash (eds), pp. 251–72.

J. P. Jones III, W. Natter and T. R. Schatzki (eds), *Postmodern contentions: epochs, politics, space* (New York and London, 1993).

M. Jones, 'The elusive reality of landscape, concepts and approaches in landscape research', *Norsk Geografisk Tidskrift* 45 (1991): 229–44.

R. Kearney, *Postnationalist Ireland. Politics, culture, philosophy* (London and New York, 1997).

G. Kearns and C. Philo (eds), *Selling places: the city and cultural capital, past and present* (Oxford, 1993).

M. Keith and S. Pile (eds), *Place and the politics of identity* (London and New York, 1993).

A. Kelly, 'Van Nost's equestrian statue of George I', *Irish Arts Review* 11 (1995): 103–7.

F. Kelly, 'The life and works of Oisín Kelly', *Irish Arts Review* 6 (1989–90): 35–8.

J. Kelly, '"The glorious and immortal": memory, commemoration and Protestant identity in Ireland, 1660–1800', *PRIA* 54, c, 2 (1992): 26–52.

J. Kelly and U. MacGearailt (eds), *Dublin and Dubliners* (Dublin, 1990).

B. P. Kennedy, *Dreams and responsibilities: the arts in independent Ireland* (Dublin, 1990).

——'The failure of the cultural republic: Ireland, 1922–1939,' *Studies* 81, 321 (1992): 14–22.

——'The Irish Free State 1922–1949: a visual perspective', in B. P. Kennedy and R. Gillespie (eds), *Ireland: art into history* (Dublin, 1994), pp. 132–54.

L. Kennedy, *Colonialism, religion and nationalism in Ireland* (Belfast, 1996).

J. Kenny, 'Portland's comprehensive plan as text: the Fred Meyer case and the politics of reading', in Barnes and Duncan (eds), pp. 176–92.

——'Climate, race and imperial authority: the symbolic landscape of the British Hill Station in India', *AAAG* 85 (1995): 694–714.

D. I. Kertzer, *Ritual, politics and power* (New Haven and London, 1988).

D. Kiberd, *Inventing Ireland: the literature of the modern nation* (London, 1996).

A. D. King, *Colonial urban development: culture, social power and environment* (London, 1976).

——*Buildings and society: essays on the social development of the built environment* (London, 1980).

——(ed.), *Representing the city* (London, 1996).

P. L. Knox, 'The social production of the built environment: architects, architecture and the post-modern city', *PHG* 11 (1987): 354–78.

——'The restless urban landscape: economic and sociocultural change and the transformation of metropolitan Washington DC', *AAAG* 81, 2 (1991): 181–209.

L. Kong, 'Negotiating conceptions of "sacred space": a case study of religious buildings in Singapore', *TIBG* 18 (1993): 342–58.

——'Political symbolism of religious building in Singapore', *E&P, D* 11 (1993): 23–45.

——'A new cultural geography? Debates about invention and reinvention', *Scottish Geographical Magazine* 113, 3 (1997): 177–85.

V. Konrad, 'Focus: nationalism in the landscape of Canada and United States', *Canadian Geographe*, Canadian Geographer 30 (1986): 167–80.

S. Kostof, *The city shaped, the city assembled*, 2 vols (London, 1991).

M. Krampen, *Meaning in the urban environment* (London, 1979).

R. Krier, *Urban space* (London, 1979).

S. Küchler, 'Landscape as memory: the mapping of process and its representation in a Melanesian society', in Bender (ed.), pp. 85–107.

B. Ladd, 'East Berlin political monuments in the late German Democratic Republic: finding a place for Marx and Engels', *Journal of Contemporary History* 37, 1 (2002): 91–104.

T. W. Laqueur, 'Memory and naming in the Great War', in Gillis (ed.), *Commemorations*, pp. 150–67.

J. J. Lee (ed.), *Ireland 1945–1970* (Dublin, 1979)

——(ed.), *Towards a sense of place* (Cork, 1985).

——*Ireland 1912–1985* (Cambridge, 1989).

L. H. Lees, 'Urban public space and imagined communities in the 1980s and 1990s', *Journal of Urban History* 20 (1994): 443–65.

H. Lefebvre, *The production of space* (Oxford, 1991).

H. Leitner and P. Kang, 'Contested urban landscapes of nationalism: the case of Taipei', *Ecumene* 6, 2 (1996): 214–33.

A. J. Lerner, 'The nineteenth century monument and the embodiment of national time', in M. Ringrose and A. J. Lerner (eds), *Reimagining the nation* (Buckingham, 1993), pp. 176–96.

S. Levinson, *Written in stone: public monuments in changing societies* (Durham, N. C., 1998).

S. J. Lewandowski, 'The built environment and cultural symbolism in post-colonial Madras', in Agnew, Mercer, and Sopher (eds), pp. 237–54.

P. Lewis, 'Axioms for reading the landscape', in Meinig (ed.), pp. 11–32.

D. Ley, 'Cultural/humanistic geography', *PHG* 9 (1985): 415–23.

——'Styles of the times: liberal and neo-conservative landscapes in inner Vancouver, 1968–86', *JHG* 13 (1987): 40–56.

——'From urban structure to urban landscape', *Urban Geography* 9, 1 (1988): 98–105.

——'Modernism, post-modernism and the struggle for place', in Agnew and Duncan (eds), pp. 44–65.

D. Ley and R. Cybriwsky, 'Urban graffiti as territorial markers', *AAAG* 64, 4 (1974): 491–505.

C. Lincoln, *The city as a work of art* (Dublin, 1992).

J. Livesy and S. Murray, 'Post-colonial theory and modern Irish culture', *Irish Historical Studies* May (1997); 452–8.

D. Lloyd, 'Introduction', in D. Lloyd, *Ireland after history* (Cork, 1999), pp. 1–18.

B. Loftus, 'Loyalist wall paintings', *Circa* 8 (1983): 10–14.

——*Mirrors: William III and modern Ireland* (Down, 1990).

H. Lorimer, 'Sites of authenticity: Scotland's new parliament and official representations of the nation', in D. C. Harvey, R. Jones, N. McInroy and C. Milligan (eds), *Celtic geographies: old culture, new times* (London, 2002) pp. 91–108.

D. Lowenthal, 'Identity, heritage, and history', in Gillis (ed), pp. 41–60.

——*The heritage crusade and the spoils of history* (London, 1996).

J. Lydon, *The making of Ireland: from ancient times to the present* (London, 1998).

F. S. L. Lyons, 'The years of readjustment', in K. B. Nowlan and T. D. Williams (eds), *Ireland in the war years and after, 1939–51* (Dublin, 1969), pp. 67–79.

——*Ireland since the famine* (London, 1973).

——*Culture and anarchy in Ireland 1890–1939* (Oxford, 1979).

D. Mac Cannell, 'The Vietnam Memorial in Washington DC', in D. MacCannell (ed.), *Empty meeting grounds* (London, 1992), pp. 280–2.

L. Mac Mathúna, 'Urban place names: streets and districts', in A. Ó Maolfabhail (ed.), *The place names of Ireland in the third millennium* (Dublin, 1992), pp. 53–70.

A. Madanipour, *Design of urban space: an inquiry into a socio-spatial process* (Chichester, 1996).

A. Martin (ed.), *W. B. Yeats: Collected Poems* (London, 1989)

D. Matless, 'Culture run riot? Work in social and cultural geography, 1994', *PHG* 19, 3 (1995): 395–403.

——'New material? Work in cultural and social geography, 1995', *PHG* 20, 3 (1996): 379–91.

I. McBride (ed.), *History and memory in Modern Ireland* (Cambridge, 2001).

L. McBride (ed.), *Images, icons and the Irish nationalist imagination* (Dublin, 1999).

E. McCann, 'Landscape, texts and the politics of planning', *E&P,D* 15 (1997): 641–61.

C. T. M'Cready, *Dublin street names* (Dublin, 1892, reprinted, 1987).

G. Macdonald, 'Indonesia's *Medan Medaka*: national identity and the built environment', *Antipode* 27, 3 (1995): 270–93.

L. McDowell, 'The transformation of cultural geography', in Gregory, Martin and Smith (eds), pp. 146–73.

MCG, *The monuments of Dublin: a poem* (Dublin and Belfast, 1865), pp. 1–20.

R. McGrath, 'Dublin panorama: an architectural review', *The Bell* 2, 5 (1941): 33–48.

J. McGregor, *Picture of Dublin* (Dublin, 1821).

E. McParland, 'A note on George II and St. Stephen's Green', *Eighteenth Century Ireland* 2 (1987): 187–95.

D. W. Meinig (ed.), *The interpretation of ordinary landscapes: geographical essays* (Oxford, 1979).

——'Geography as an art', *TIBG* 8 (1983): 314–28.

S. Michalski, *Public monuments: art in political bondage, 1870–1997* (London, 1998)

D. Mitchell, 'There's no such thing as culture: towards a reconceptualisation of the idea of culture in geography', *TIBG* 20 (1995): 102–16.

——'The end of public space? People's Park, definitions of the public, and democracy', *AAAG* 85, 1 (1995): 108–33.

——'Explanation in cultural geography: a reply to Cosgrove, Jackson and the Duncan's', *TIBG* 21, 3 (1996): 580–2.

——'Sticks and Stones: the work of landscape', *Professional Geographer* 48 (1996): 94–6

——*Cultural geography: a critical introduction* (Oxford, 2000), p. xiv.

W. J. T. Mitchell, *Iconology, image, text, ideology* (Chicago, 1986).

——(ed.), *Art and the public sphere* (Chicago, 1990)

——(ed.), *Landscape and power* (Chicago, 1994).

J. Monk, 'Gender in the landscape: expressions of power and meaning', in F. Gale and K. Anderson (eds), *Inventing places: studies in cultural geography* (Melbourne, 1992), pp. 123–38.

J. Morisy, *A wreath for the O'Brien statue* (Dublin, 1870).

A. E. J. Morris, *History of urban form before the industrial revolutions* (London, 1994).

G. Mosse, *The nationalisation of the masses: political symbolism and mass movements in Germany from the Napoleonic wars to the Third Reich* (New York, 1975).

——*Masses and man: nationalist and fascist perceptions of reality* (New York, 1980).

——*Fallen soldiers: reshaping the memory of the world wars* (Oxford, 1990).

R. Muir, 'Landscape: a wasted legacy', *Area* 30, 3 (1998): 263–71.

L. Mumford, *The city in history* (Harmondsworth, 1961).

——*The culture of cities* (New York, 1983).

J. A. Murphy, *Ireland in the twentieth century* (Dublin, 1989).

P. Murphy, 'Terence Farrell, sculptor', *Irish Arts Review* 8 (1991–2): 73–9.

——'Thomas Farrell sculptor', *Irish Arts Review* 9 (1992–3): 196–207.

——'The politics of the Irish street monument', *Irish Arts Review* 10 (1994): 202–8.

——'John Henry Foley's O'Connell monument', *Irish Arts Review* 11 (1995): 155–6.

——'The O'Connell monument in Dublin: the political and artistic context of a public sculpture', *Apollo* March (1996): 22–6.

——'The quare on the square: a statue of Oscar Wilde for Dublin', in J. McCormack (ed.), *Wilde the Irishman* (London and New Haven, 1998), pp. 127–39.

H. Murray, 'The pillar project', *Irish Architect*, Nov./Dec. (1988): 20–1.

G. A. Myers, 'Making the socialist city of Zanzibar', *Geographical Review* 84, 2 (1994): 451–62.

——'Naming and placing the other: power and the urban landscape in Zanzibar', *Tijdschrift voor Economische en Sociale Geografie* 87, 3 (1996): 237–46.

C. Nash, 'Irish placenames: post-colonial locations,' *TIBG*, 24, 3 (1999): 457–80.

W. Natter and J. P. Jones, 'Signposts toward a poststructuralist geography', in Jones, Natter and Schatzki (eds), pp. 165–203.

C. Nelson, P. Treichler and L. Grossberg, 'Cultural studies: an introduction', in L. Grossberg, C. Nelson and P. Treichler (eds), *Cultural studies* (New York, 1992), pp. 1–16.

P. Nora, '"Between memory and history", les lieux des memoires', *Representations* 26 (1989): 10–18.

M. North, 'The public as sculpture: from heavenly city to mass ornament', in W. J. T. Mitchell (ed.), *Art and the public sphere* (Chicago, 1990), pp. 9–28.

J. V. O'Brien, *Dear dirty Dublin: a city in distress, 1899–1916* (Berkeley and Los Angeles, 1982).

B. Ó Conaire (ed.), *Douglas Hyde, language, love and lyrics: essays and lectures* (Dublin, 1986)

E. E. O'Donnell, *The annals of Dublin, fair city* (Dublin, 1987).

D. O'Donovan and A. Powers, *God's architect: a life of Raymond McGrath* (Bray, 1995).

F. O' Dwyer, *Lost Dublin* (Dublin, 1981).

J. C. O' Hanlon, *Report of the O'Connell monument committee* (Dublin, 1888).

T. O'Keefe, 'The art and politics of the Parnell monument', *Éire–Ireland* 19 (1984): 6–25.

——'The 1898 efforts to celebrate the United Irishmen', *Éire–Ireland* 23 (1988): 51–73.

——'Who fears to speak of '98? The rhetoric and rituals of the United Irishmen centennial, 1898', *Éire/Ireland* 27 (1992): 67–91.

S. O'Laoire, 'Building on the edge of Europe', in J. Graby (ed.), *Building on the edge of Europe/Construire à la frange de l'Europe: a survey of contemporary architecture in Ireland embracing history, town and country* (Cork, 1996), pp. 118–67.

A. Ó Maolfabhail (ed.), *The placenames of Ireland in the third millennium* (Dublin, 1992).

Office of Public Works, 'Memorial to Thomas Davis', *Oibre* 2 (1965).

——'Links with the war years', *Oibre* 4 (1967).

——'The Davis Memorial', *Oibre* 4 (1967).

——'Tone, Emmet, Yeats: three memorials in St. Stephen's Green', *Oibre* 6 (1968).

——'Garden of Remembrance Sculptural Group', *Oibre* 10 (1973).

——'The Wellington Monument', *Oibre* 15 (1978).

J. Olley, 'The theatre of the city of Dublin, 1991', *Irish Arts Review* 9 (1993): 70–8.

——'Weaving the fabric of the city. Urban design and Dublin', in Becker, Olley and Wang (eds), pp. 42–7.

——'Airport terminal building', in Becker, Olley and Wang (eds), pp. 110–13.

D. J. Olsen, *The city as a work of art*. (London, 1986).

K. Olwig, 'Recovering the substantive nature of landscape', *AAAG* 86, 4 (1996): 630–3.

J. O'Regan, *Michael Scott, 1905–1989* (Dublin, 1993).

——(ed.), *A monument in the city* (Dublin, 1998).

M. Ó Riain, 'Nelson's Pillar: a controversy that ran and ran', *History Ireland* 6, 4 (1998): 21–5.

B. S. Osborne, 'The iconography of nationhood in Canadian Art', in Cosgrove and Daniels (eds), pp. 162–78.

——'Constructing landscapes of power: the George Étienne Cartier monument, Montreal', *JHG* 24, 4 (1998): 431–58.

——'Warscapes, landscapes, inscapes: France, War and Canadian national identity', in I. S. Black and R. A. Butlin (eds), *Place, culture and identity* (Quebec, 2001), pp. 311–34.

——'Interpreting a nation's identity: artists as creators of national consciousness', in A. R. H. Baker and G. Biger (eds), *Ideology and landscape in historical perspective* (Cambridge, 1992), pp. 230–54.

W. N. Osborough, *Law and the emergence of modern Dublin* (Dublin, 1996).

J. O'Sheehan and E. de Barra, *Oispidéil na hÉireann: Ireland's hospitals, 1930–1955* (Dublin, 1956).

S. O'Toole, *Collaboration: the pillar project* (Dublin, 1988).

G. Owens, 'Nationalist monuments in Ireland, 1870–1914: symbolism and ritual', in R. Gillespie and B. P. Kennedy (eds), *Ireland: art into history* (Dublin, 1994), pp. 103–17.

K. Palonen 'Reading street names politically', in K. Palonen and T. Parviko (eds), *Reading the political: exploring the margins of politics* (Tampere, 1993), pp. 103–21.

E. Panofsky, *Studies in iconology: humanistic themes in the art of the renaissance* (Oxford, 1939).

——'Iconography and iconology: an introduction to the study of Renaissance art', in E. Panofsky, *Meaning the visual arts* (Harmondsworth, 1970), pp. 51–81.

J. J. Parsons, 'Carl Ortwin Sauer, 1889–1975', obituary, *Geographical Review* 66 (1976): 83–9.

——'The later Sauer years', *AAAG* 69 (1979): 9–24.

——'Cultural geography at work', in Foote, Hughill, Mathewson, and Smith (eds), pp. 281–8.

S. Paseta, '1798 in 1898: the politics of commemoration', *Irish Review* 22 (1998): 46–53.

R. Peet, 'Discursive idealism in the 'landscape as text school', *Professional Geographer*, 48, 1 (1996): 96–8.

——'A sign taken for history: Daniel Shays' memorial in Petersham, Massachusetts', *AAAG* 86, 1 (1996): 21–43.

C. B. Peterson, 'The nature of Soviet placenames', *Names* 32, 4 (1977): 435–42.

C. Philo, 'New directions in cultural geography: a conference of the social geography study group of the IBG, UCL, September 1–3, 1987', *JHG* 14 (1988): 178–81.

——(ed.), *New worlds, new worlds: reconceptualising social and cultural geography*. Departments of Geography, University of Edinburgh and St David's University College, Lampeter, Wales for the social and cultural geography group of the IBG.

P. Pickering, *Pictorial record of centenary of Thomas Davis and Young Ireland* (Dublin, 1945).

G. K. Piehler, 'The war dead and the gold star: American commemoration of the First World War', in Gillis (ed.), pp. 168–85.

S. Pile, 'Practising interpretative geography', *TIBG* 16 (1991): 458–69.

H. Potterton, 'Dublin's vanishing monuments', *Country Life* May (1974): 1304–5.

——*The O'Connell monument* (Cork, 1973).

A. Pred, 'Capitalisms, crises, and cultures II: notes on local transformation and everyday struggles', in A. Pred and M. Watts, *Reworking modernity: capitalisms and symbolic discontent* (New Brunswick, NJ, 1992), pp. 106–17.

M. Price and M. Lewis, 'The reinvention of cultural geography', *AAAG* 83, 1 (1993): 1–17.

——'Reply: on reading cultural geography', *AAAG* 83 (1993): 515–22.

J. Prunty, 'From city slums to city sprawl: Dublin from 1800 to the present', in H. B. Clarke (ed.), *Irish cities* (Dublin and Cork, 1995), pp. 109–22.

——*Dublin slums* (Dublin, 1998).

P. J. Raivo, 'The limits of tolerance: the orthodox milieu as an element in the Finnish cultural landscape, 1917–1939', *JHG* 23, 2 (1997): 327–39.

A. Rappoport, *The meaning of the built environment* (Beverly Hills, 1982).

E. Renan, 'What is a nation?', trans. M. Thom, in H. Bhabha (ed.), *Nation and narration* (London, 1990), pp. 8–22.

Review of J. T. Gilbert, *History of the city of Dublin*, *Dublin University Magazine* March (1856): 320–43.

J. J. Robinson, 'Comments on the foregoing article, No. 3', *Studies* (1942): 164–6.

A. Rolfe and R. Ryan, *The Department of Industry and Commerce, Kildare Street, Dublin* (Dublin, 1992).

G. Rose, 'Discourse analysis 1: text, intertextuality, context', in G. Rose, *Visual methodologies* (London, 2001), pp. 135–63

S. Rothery, 'Parnell monument: Ireland and the Beaux Arts', *Irish Arts Review* 4, 1 (1987): 55–7.

——*Ireland and the new architecture, 1900–1940* (Dublin, 1991).

R. Rottenberg and G. McDonagh (eds), *The cultural meaning of urban space* (Connecticut, 1993).

L. B. Rowntree and M. W. Conkey, 'Symbolism and the cultural landscape', *AAAG* 70, 4 (1980): 459–74.

E. Said, 'Narrative, geography and interpretation', *New Left Review* 180 (1994): 81–97.

——'Invention, memory, and place', *Critical inquiry* 26 (2000): 175–92.

M. S. Samuels and C. M. Samuels, 'Beijing and the power of place in modern China', in Agnew and Duncan (eds), pp. 202–27.

M. S. Samuels, 'The biography of landscape', in Meinig (ed.), pp. 51–88.

R. Samuel, *Theatres of memory; vol. 1, past and present in contemporary culture* (London, 1994)

L. Sandercock, *Towards cosmopolis* (New York, 1988), p. 207.

C. O. Sauer, 'The morphology of landscape', *University of California Publications in Geography* 2 (1925): 19–54.

S. Schama, *Landscape and memory* (London, 1995).

R. H. Schein, 'The place of landscape: a conceptual framework for interpreting an American scene', *AAAG* 87 (1997): 660–80.

C. Schorske, *Fin de Siècle Vienna: politics and culture* (New York, 1981).

W. A. Scott, 'Reconstruction of O'Connell Street Dublin: a note', *Studies*, Nov. (1916): 165.

J. C. Scruggs and J. Swerdlow, *To heal a nation: the Vietnam veteran's memorial* (New York, 1985).

Sculptors' Society of Ireland, *Contemporary sculpture in Dublin: a walker's guide* (Dublin, 1991).

S. Seymour, 'Historical geographies of landscape', in Graham and Nash (eds), *Modern historical geographies* (Harlow, 2000), pp. 193–217.

M. Shaffrey, 'Sackville Street/O'Connell Street', *Irish Arts Review* 5 (1988): 144–56.

H. Shaw, *New city pictorial directory* (Dublin, 1850).

B. J. Shaw and R. Jones (eds), *Contested urban heritage: voices from the periphery* (Aldershot, 1997).

N. Sheaff, 'The Shamrock building', *Irish Arts Review* 1 (1984): 26–9.

J. Sheehy, *The rediscovery of Ireland's Celtic past, 1830–1930* (London, 1980).

E. Sheridan, 'Designing the capital city: Dublin ca. 1660–1810', in Brady and Simms (eds), pp. 66–135.

J. R. Short, *The urban order: an introduction to cities, culture, and power* (London, 1996).

P. Shurmer-Smith and K. Hannam, 'Monuments and spectacles', in P. Shurmer-Smith and K. Hannam, *Worlds of desire, realms of power* (London, 1994), pp. 198–214.

P. Shurmer-Smith (ed.), *Doing cultural geography* (London, 2001)

D. Sidorov, 'National monumentalisation and the politics of scale', *AAAG* 90, 3 (2000): 548–72.

J. Smith, (1993), 'The lie that binds: destabilising the text of landscape', in Duncan and Ley (eds), pp. 78–95.

G. Smyth, *Space and the Irish cultural imagination* (Basingstoke, 2001).

W. Smyth, 'Explorations of place', in J. J. Lee (ed.), *Towards a sense of place* (Cork, 1985).

E. W. Soja, *Postmodern geographies: the reassertion of space in critical social theory* (London, 1989).

——*Thirdspace: a journey through Los Angeles and other read and imagined places* (Oxford, 1995).

W. Stokes, *The Life and labours in art and archaeology of George Petrie* (London, 1868).

R. W. Stump, 'Toponymic commemoration of national figures: the case of Kennedy and King', *Names* 36 (1988): 203–16.

W. M. Thackeray, *The Irish sketchbook* (Dublin, 1842, reprinted Gloucester, 1990).

Yi-Fu Tuan, 'Space and place: humanistic perspectives', *PHG* 6 (1974): 213–52.

——*Topophilia: a study of environmental perception, attitudes and values* (Englewood Cliffs, 1974).

——'Place: an experiential perspective', *Geographical Review* 65 (1975): 151–65.

J. E. Tunbridge, 'The question of heritage in European cultural conflict', in Graham (ed.), pp. 236–60.

J. E. Tunbridge and G. J. Ashworth, *Dissonant heritage: the management of the past as a resource in conflict* (Chichester, 1996).

J. Turpin, 'Oliver Sheppard's 1798 memorials', *Irish Arts Review* 7 (1991): 71–80.

——'Cúchulainn lives on', *Circa* 69 (1994): 26–31.

T. Unwin, *The place of geography* (London, 1992).

——(ed.), *A European geography* (London, 1998).

L. J. Vale, *Architecture, power and national identity* (New Haven and London, 1992).

P. L. Wagner and M. W. Mikesell, *Readings in cultural geography* (Chicago, 1962).

D. Walker, *Michael Scott architect* (Kinsale, 1995).

S. Walker, 'Architecture in Ireland, 1940–1975', in Becker, Olley and Wang (eds), pp. 23–9.

J. R. Walton, 'How realist can you get?' *Professional Geographer* 47, 1 (1995): 61–5.

——'Bridging the divide – a reply to Mitchell and Peet', *Professional Geographer* 48, 1 (1996): 98–100.

J. Warburton, J. Whitelaw and R. A. Walsh, *A history of the city of Dublin*, 2 vols (London, 1818).

G. K. Whammond, *The exhibition guide through Dublin, containing a description of the public buildings, cathedrals, churches and chapels* (Dublin, 1853).

K. Whelan, *The tree of liberty: radicalism, Catholicism and the construction of Irish identity* (Cork, 1996).

Y. Whelan, 'Monuments, power and contested space: the iconography of Sackville St (O'Connell Street) before independence (1922) *Irish Geography* 34, 1 (2001): 11–33.

——'Symbolising the state: the iconography of O'Connell Street after independence (1922)', *Irish Geography* 34, 2 (2001): 135–56.

——'The construction and destruction of a colonial landscape: monuments to British monarchs in Dublin before and after independence', *JHG* 28, 4 (2002): 508–33.

M. Williams, 'Historical geography and the concept of landscape', *JHG* 15 (1989): 92–104.

H. Winchester, 'The construction and deconstruction of women's roles in the urban landscape', in Anderson and Gale (eds), *Inventing places*, pp. 139–56.

D. S. Winton, and N. G. Kotler, *The Statue of Liberty revisited: making a universal symbol* (Washington and London, 1994).

C. W. J. Withers, 'How Scotland came to know itself: geography, national identity and the making of a nation, 1680–1790', *JHG* 21, 4 (1995): 371–97.

——'Place, memory, monument: memorialising the past in contemporary Highland Scotland', *Ecumene* 3, 3 (1996): 325–44.

P. Woolf, 'Symbol of the Second Empire: cultural politics and the Paris Opera House', in Cosgrove and Daniels (eds), pp. 214–35.

B. S. Yeoh, 'Street names in colonial Singapore', *Geographical Review* 82, 3 (1992): 313–22.

——'The politics of space: changing discourses on Chinese burial grounds in post-war Singapore', *JHG* 21, 2 (1995): 184–201.

——'Street-naming and nation-building: toponymic insciptions of nationhood in Singapore', *Area* 28, 3 (1996): 298–307.

——'Historical geographies of the colonised world', in Graham and Nash, (eds), pp. 146–66

——'Postcolonial cities', *PHG* 25, 3 (2001): 456–68.

A. Yerolympos, 'A new city for a new state: city planning and the formation of national identity in the Balkans', *Planning Perspectives* 8 (1993): 233–57.

E. Young, 'The biography of a memorial icon: Nathan Rapoport's Warsaw Ghetto monument', *Representations* 26 (1989): 69–106.

S. Zukin, *Landscapes of power: from Detroit to Disney World* (Berkeley, 1991).

——'Postmodern urban landscapes: mapping culture and power', in Jones, Natter and Schatzki (eds), pp. 223–45.

Index

Note: Page references in italics denote illustrations

297

Index